MACHINERY
CONDITION
MONITORING
PRINCIPLES AND PRACTICES

MACHINERY CONDITION MONITORING

PRINCIPLES AND PRACTICES

AMIYA R. MOHANTY

Indian Institute of Technology
Mechanical Engineering Department
Kharagpur, West Bengal, India

CRC Press
Taylor & Francis Group
Boca Raton London New York

CRC Press is an imprint of the
Taylor & Francis Group, an **informa** business

CRC Press
Taylor & Francis Group
6000 Broken Sound Parkway NW, Suite 300
Boca Raton, FL 33487-2742

First issued in paperback 2017

© 2015 by Taylor & Francis Group, LLC
CRC Press is an imprint of Taylor & Francis Group, an Informa business

No claim to original U.S. Government works

ISBN-13: 978-1-4665-9304-6 (hbk)
ISBN-13: 978-1-138-74825-5 (pbk)

Dedicated to the memory of my father.

Professor Aswini Kumar Mohanty

(1941–1995)

Contents

Preface

While teaching the subject of machinery fault diagnostics and signal processing, and conducting short courses for industry professionals for more than fifteen years to the senior year and graduate students at the Indian Institute of Technology, Kharagpur, I have always had a problem locating a single book on the subject. Though there are excellent research journals and trade magazines on the subject, I have failed to locate a book without bias toward the equipment used or process followed for detecting faults in machines in general, and rotating machines in particular. With this preamble I began to write this book, drawing on my experience with machinery condition monitoring while consulting for many industries in India and abroad over the last two decades.

In many of my fault-detection exercises in the industries where I have consulted, I have come across technicians and engineers who wonder what fast Fourier transform (FFT) is, and I have been provided with a table full of numbers and asked to find the fault in the machines. Many engineers in the field learn this subject from the technical brochure of the equipment they use, without understanding their limitations. They feel that since they have a piece of costly equipment at their disposal, they can detect and diagnose the fault in any machine. One of the purposes of this book is to eliminate such myths. No offense intended. There are many commercial software packages available in the market to do diagnostics, but with an understanding of the subject, an engineer can better appreciate the diagnostic routines, because garbage in is garbage out.

This book can be used as a textbook in universities for a forty-hour lecture class. Much of the material in this book has been time tested in my class at the Indian Institute of Technology Kharagpur. I have found MATLAB® to be a very easy way to introduce students to the basics of signal processing, a very important component in machinery condition monitoring. A few MATLAB codes for signal processing are provided in this book.

MATLAB is a registered trademark of The MathWorks, Inc. For product information, please contact

The MathWorks, Inc.
3 Apple Hill Drive
Natick, MA 01760-2098 USA
Tel: 508 647 7000
Fax: 508-647-7001
E-mail: info@mathworks.com
Web: www.mathworks.com

Various techniques of machinery maintenance are available today, and after a brief overview of the maintenance techniques in Chapter 2, I focus on the fundamentals of machinery vibration and rotor dynamics in Chapters 3 and 4. A majority of the condition monitoring techniques in the world involve monitoring machine vibrations. Signal processing is a very important component in diagnosing the fault in a machine. Thus Chapters 5 and 6 are focused on the basics of signal processing and instrumentation, which are essential for monitoring the health of machines. Discussions of the requirements of vibration monitoring and noise monitoring are presented in Chapters 7 and 8. Lately, motor current signature analysis has developed into one of the major techniques of fault detection in machines. For example, how does one monitor the condition of a submersible pump buried in the ground, where a vibration transducer cannot be mounted? Further, electric motors are used in a majority of situations in the industry as prime movers, so monitoring their condition is of paramount importance. Chapter 9 focuses on electrical machinery faults. Infrared temperature detection has emerged as a cheap and convenient way to detect hot spots, so Chapter 10 describes the thermography for condition monitoring. Chapters 11 and 12 discuss the techniques of wear debris analysis and some of the nondestructive test (NDT) techniques for condition monitoring like ultrasonics, radiography, and so on. In manufacturing industries, the information regarding a cutting tool, in particular when the tool should optimally be changed, is a matter of concern. Chapter 13 discusses machine tool condition monitoring. In spite of our best maintenance efforts, machines do fail for reasons beyond our control. After a machine fails, we are called to do a postmortem. So, I have added a chapter on engineering failure analysis (Chapter 14). In Chapter 15 I discuss several case studies, mostly on failure analysis, from my consulting experience. After going through all the earlier chapters, I sincerely hope that readers will be able to diagnose the faults in machines on their own. I have done my job if readers can do that. This book concludes with a bibliography for those interested in learning more about the subject, and a set of appendices on standards, balancing, misalignment techniques, and the use of MATLAB®.

I hope this book will be on the desk of every practicing maintenance engineer, student studying the subject, and my colleagues from academia who are teaching and doing research in the areas of machinery condition monitoring.

A. R. Mohanty
Kharagpur
June 2014

Acknowledgments

Writing a book is not an easy thing. At the outset I must thank all of the students in my machinery condition monitoring and signal processing class for their intriguing questions, which initially led me to believe that a single source book on this subject needed to be written—and the sooner the better. A course on machinery condition monitoring cannot be taught from research papers and fundamental engineering books. A lot of field experience needs to be brought into the classroom. I thus thank my many industry sponsors in India and some in the United States, who gave me several opportunities to do field measurements and then provide solutions to fix problems. Such experience was very helpful to me as a consultant and then as an author of this book. I have tried to bring many such field topics into this book.

Of course, my training as a mechanical engineer, with close to thirty years of experience in machinery condition monitoring, would not have been possible without the care and guidance of my teachers, mentors, and colleagues. I wish to acknowledge my undergraduate engineering teachers at NIT Rourkela, Professors T. N. Subramanian and A. K. Behera, who introduced me to vibrations. I thank Professor Amalendu Mukherjee, my teacher and master's guide at the Indian Institute of Technology Kharagpur who, on the very first day I met him, introduced me to fast Fourier transform (FFT), and of course then many other rudiments and aspects of data acquisition, signal processing, and controls. My PhD supervisor Professor Andy Seybert at the University of Kentucky taught me, among many other things that I will always cherish, how to do correct measurements, and actually taught me the rudiments of acoustics and noise control and how to be a successful engineering consultant. I thank Professor Seybert for instilling in me the confidence to do experiments. Professor Bob Bernhard, my postdoctoral guide at Purdue, introduced me to the noise vibration harshness (NVH) chassis dynamometer at Herrick Labs, where we did sponsored research on active control of road noise.

I have also had a few mentors in the industries where I have worked. Mr. R. A. Noras and Mr. Amitabh Nath at Larsen Toubro Limited, Mumbai, gave me many of my first opportunities to conduct machinery field vibration measurements, machinery failure analyses, and experimental modal analyses. My experience at Ford Motor Company was very rewarding and allowed me to do many vehicle noise and vibration measurements at the Dearborn Proving Ground and do computer-aided engineering (CAE) simulations. For these experiences, I owe a debt of gratitude to my supervisors, managers, and colleagues at Ford, especially Dr. Farhang Aslani, Dr. Randy Visantainer, Dr. P. Suruli-Naryansami, Mr. Tom Martin, Mr. Bruce Bonhard, Mr. Barry St. Pierre, and Ms. Stacey McCreery.

This book was made possible by the research in machinery condition monitoring being done by my group at the Indian Institute of Technology Kharagpur. A few of the examples in this book are based on the results of my past and present PhD students. I wish to acknowledge my past students, Dr. S. Prabhakar, Dr. C. Kar, Dr. A. K. Jalan, Dr. (Cdr) V. K. Rai, and Dr. P. K. Pradhan. I also want to thank my present students, S. Fatima, S. K. Roy, T. Bose, P. Bansod, Marxim, and Bhartendra for their help in conducting a few experiments on the rigs.

While working as a faculty member at the Indian Institute of Technology, I get to interact with other academic and nonacademic institutes throughout the world. I would like to acknowledge my interactions with Professor M. L. Munjal from the Indian Institute of Science Bangalore; Professor S. Narayanan, Professor C. Sujatha, Professor P. Chandramouli, Professor A. S. Sekhar, and Professor B. V. A. Rao from the Indian Institute of Technology Madras; Professor D. N. Manik from the Indian Institute of Technology Bombay; Professor N. Tandon from the Indian Institute of Technology Delhi; Professor N. Tiwari from the Indian Institute of Technology Kanpur; Professor R. Tiwari from the Indian Institute of Technology Guwahati; Professors A. B. Chattopadhyay, C. S. Kumar, Siddhartha Das, Karabi Das, A. Routray, A. Patra, S. Mukhopadhyay, S. Paul, V. N. A. Naikan, and K. Pathak from the Indian Institute of Technology Kharagpur; and Dr. V. Bhujanga Rao from the Defence Research Development Organization and the founder president of the Condition Monitoring Society of India.

I acknowledge international interactions with Professor D. Herrin and Professor T. W. Wu of the University of Kentucky; Professor W. L. Li of Wayne State University; Professor Yoke San Wong and Professor H. S. Hong of the National University of Singapore; Professor M. Pekpe from University of Lille 1, France; Professor Paul Jennings of the Warwick Manufacturing Group; and Professor Aditya Parida of Lulea University, Sweden, for the many collaborative projects and student exchanges I have had with them.

Of course, I thank all the members of my family and close friends, who have been very supportive during my book-writing sessions. I thank my wife, Bithika, and my two sons Saurav and Siddarth, who kept reminding me of the target at hand and the much awaited vacation after I completed the manuscript. I must thank Dr. Gagandeep Singh, acquisitions editor at CRC Press, for the timely reminders and bearing with me when I exceeded the deadlines. This book just would not have been possible without my interactions with all the persons I have mentioned above. I owe each one of them my sincere thanks and gratitude. Please excuse me if I inadvertently missed anyone.

About the Author

Amiya Ranjan Mohanty has been a faculty member at the Indian Institute of Technology Kharagpur, India, since 1996, where he is currently a professor of Mechanical engineering. Professor Mohanty has a B.ScEngg (Hons) degree in mechanical engineering from the National Institute of Technology, Rourkela (formerly the Regional Engineering College). He holds a master's degree in machine design specialization from the Indian Institute of Technology, Kharagpur, and a PhD in the area of noise control from the University of Kentucky in the United States.

At the Indian Institute of Technology, Kharagpur, he established a state-of-the-art laboratory for acoustics and condition monitoring. He teaches courses in machine design, instrumentation, automobile engineering, acoustics and noise control, machinery fault diagnosis, and signal processing. He regularly conducts short-term courses in the areas of noise control and condition monitoring. He has been a consultant to more than 50 companies in the areas of noise control and machinery condition monitoring.

Professor Mohanty has worked in the R&D divisions of Larsen Toubro Limited, Mumbai, and Ford Motor Company at Dearborn, Michigan, in the United States. Professor Mohanty has collaborated in research on noise control and machinery condition monitoring with many leading universities around the world in the United States, France, Sweden, the United Kingdom, and Singapore, and was a postdoctoral Fellow at the Ray W. Herrick Labs at Purdue University in the United States.

Professor Mohanty is a Fellow of the Acoustical Society of India and a member of ASME and SAE. He received the Chancellor's Award for outstanding teaching at the University of Kentucky, the Rais Ahmed Memorial Award of the Acoustical Society of India, and the best research paper in underwater acoustics from the Acoustical Society of India. He has published more than 100 journal articles in the areas of noise control and machinery condition monitoring. His research projects are sponsored by several private and government agencies.

1

Introduction

1.1 Machinery Condition Monitoring

Different types of machinery are present in industry. The majority of these machines have rotating components. For an industrial plant to have high production rates, the machines must perform as per their designed specifications and installed capacity. In order to ensure that these machines run without any significant downtime, the machine must be in proper condition. The maintenance personnel in a plant take necessary steps to ensure that the machines run for their designed life without any significant failure.

Machinery condition monitoring deals with the maintenance aspects of these machines based on the present and past condition of the machine. In order to know the machine's condition, sensors are installed around the machine so that relevant information about the machine's health condition can be collected, analyzed, and decisions made regarding the appropriate maintenance or corrective actions to be taken so that the machine is able to perform as per its original design objectives. In a plant without a proper protocol in place for machine maintenance, the end result could be an eventual loss for the plant. Machinery maintenance is dependent on the type of machine, the severity of the defects in the machines, and the downstream consequences they may have on the overall operation of the plant. The different aspects of maintenance will be dealt with in the next chapter. Machinery condition maintenance is a scheme in which appropriate maintenance is done based on the condition of the machine.

During operation, machines give out information or signals in the form of noise, vibration, temperature, lubricating oil condition, quality and quantity of the motor current drawn, and the like. These signals from the machine are acquired by installing transducers to measure the mechanical parameters of the machine. The signals thus obtained are usually analog and they exist at all times. In order to create meaningful information from these signals, the signals are converted into the digital domain by analog-to-digital converters. The discrete digital data corresponding to the analog signal thus obtained is analyzed on computers. Software is available to efficiently store and handle the large digital data collected from machines. These data can be used in algorithms developed for fault detection in machines. Once the fault in machines has been diagnosed, corrective measures can be initiated so that the machine has a long useful life, and the plant has high productivity.

1.2 Present Status

As was mentioned in the previous section, a few of the important aspects of machinery condition monitoring are transducers, instrumentation, signal analysis software, and a decision-making system. Developments are continually being made in all the above aspects. In a majority of the machines throughout the world, the machine's condition is monitored through vibration measurements. In most the cases, the vibrations are measured using contact-type piezoelectric accelerometers. For noncontact vibration measurements, laser-based systems are available that can measure both transverse and rotational vibration of a rotating machine. The wear and debris deposited in the lubricating oil of a machine are analyzed for their chemical composition along with an oil analysis.

Wear debris and oil analysis are usually done in dedicated third-party laboratories, though some in situ analysis is done with handheld particle count meters to have an estimate of the size and concentration of the wear debris particles suspended in the lubricating oil. Motor current signature analysis is an established technique to determine faults in the electrical motor driving the mechanical machine. However, recently this has been extended to also detect defects in mechanical units like pumps and gearboxes by analyzing the quality of the current drawn by the electric motor.

Many nondestructive techniques (NDT) have also been made available to determine internal defects in machine components like internal cracks, weld defects, surface cracks not visible to the naked eye, and the like. The phased-array technique is one such recent development in ultrasonic testing, where a three-dimensional view of the inside of a component can be obtained in one surface scan of the component. Such scanning techniques are quick and time saving. Another technique that is popular recently is the use of infrared thermal imaging to detect regions of high temperature in a machine component caused by high heat generation due to a defect. Such techniques are quick to pinpoint the source of a defect.

Along with transducers, many developments have been made in instrumentation and signal conditioning. Smart sensors with embedded electronics are available, where the sensitivity settings along with the identification serial number are stored in the sensor itself. This makes it convenient for diagnostic personnel because the proper sensitivity does not have to be input to the signal analysis system. Sensors are available that warn the computer-based system if there is a cable fault. Due to developments in electronics, the systems have become small and have less weight and are thus easily portable. The battery power source for the instrumentation system has also been developed have a higher power-to-weight ratio, such as lithium-ion cells. Data acquisition systems have become very portable and are usually personal computer–based systems with a universal serial bus (USB) interface of the plug-and-play type. Digital signal processing algorithms are available that are fast, and can even determine the signal characteristics of a very short-duration acquired signal from the rotating machine. Efficient database management software is available that integrates the maintenance aspects of the plant with the production rate, raw material inventory, warehousing, and delivery schedules. Such enterprise resource planning (ERP) software is becoming quite popular in modern industries, since all the required information is available in one place for any interested plant personnel.

1.3 Fault Prognosis

Faults in machines can be diagnosed by analyzing the signals obtained from transducers installed around the machines. The fault diagnosis depends on the characteristics of the signals, and incipient faults can also be detected with efficient algorithms. Once the faults have been detected and diagnosed, the next question is how long the machine will last in the present condition or what is the remaining useful life (RUL) of the machine under observation. Many deterministic and stochastic approaches are available for predicting the RUL of a machine.

1.4 Future Needs

New techniques for fault diagnosis in machines are always emerging with the advancement of technology. There has been improvement in all aspects of condition monitoring, such as transducers and instrumentation, digital signal processing, and decision making, although with the complexities of machines increasing day by day and their size reducing, their maintenance is also a challenge. The use of wireless systems for data transfer from the measurement location to the analysis station is preferred because it eliminates the need for long cables from the measurement location to the analysis station, though the current available wireless transfer speeds, limit high-speed data acquisition of dynamic vibration signals.

Sensors should be fault tolerant and have provision for automatic self-calibration. Using embedded electronics, the sensor, signal conditioner, and analysis and decision system should be one standalone dedicated miniature unit. Motor current signature analysis can be used to monitor machines driven by electric motors at remote locations, since access to the machine is not necessary as long as the current carrying conductor to the electric motor can be accessed for current signature analysis. Many of the above are already available in many commercial condition monitoring systems and improvements are being made throughout the world by researchers and developers. Many of these issues are presented in this book.

2

Principles of Maintenance

2.1 Introduction

A machine that has been designed and manufactured to perform a certain function, is expected to do so when installed in a plant for its designed life span. However, for reasons beyond one's control, such a machine may fail to do so for several reasons. Some of the reasons could be a faulty design of the machine, inferior material and workmanship, incorrect installation and wrong operational procedure, among many others. However, in a plant where the output could be finished goods, the failure of the installed machine would lead to a loss of sales and loss in the earnings made by the plant. So, if care is not taken of the machine to avoid failures, a plant owner could incur serious financial loss and lead to bankruptcy. The machine thus has to be maintained to avoid such failures. All over the world, plant operators adopt three different types of maintenance techniques for machines, known as the reactive maintenance, preventive maintenance, and predictive maintenance. The benefits of planned maintenance are as follows:

- Eliminate unnecessary maintenance
- Reduce rework costs
- Reduce lost production caused by failures
- Reduce repair parts inventory
- Increase process efficiency
- Improve product quality
- Extend the operating life of plant systems
- Increase production capacity
- Reduce overall maintenance costs
- Increase overall profit

In this chapter, a brief overview of these maintenance techniques is provided along with a procedure known as *failure modes effects and criticality analysis* (FMECA), which suggests a scientific way to decide about the type of maintenance to be adopted for a particular machine in a plant.

2.2 Reactive Maintenance

As the term implies, *reactive maintenance* means that we react to a need for maintenance of the machine, because the machine has failed completely. Another name for reactive maintenance is *breakdown maintenance*. Obviously, in this type of maintenance, no maintenance is done on the machine, and only when the machine fails it is replaced by an entirely new machine. So in a plant, machines that are very critical and expensive obviously cannot be left to fail by performing reactive maintenance. Usually, the less expensive and noncritical machines can be good candidates for reactive maintenance. For example, in a steel plant one obviously cannot afford to have the blast furnace under a reactive maintenance program, but perhaps a water cooler in the workers' cafeteria may be a candidate for reactive maintenance. Following are the attributes of a reactive maintenance program:

- High expenses involved
- High spare parts inventory cost
- High overtime labor costs
- High machine downtime
- Low production availability

2.3 Preventive Maintenance

In *preventive maintenance*, the maintenance on a particular machine is done in a regular periodic manner at a fixed frequency. This type of maintenance is also known as *periodic maintenance*. Normally, based on the designer's input, the original equipment manufacturer (OEM) provides a schedule for the periodic maintenance to be done on the machine. The guiding principle behind this maintenance technique is that the machine is up to performing its desired function at any time, since it has been "well" maintained. While doing preventive maintenance on a machine, sometimes some components are replaced that are otherwise functional and do not warrant replacement. For example, when we purchase a new vehicle, we all take our vehicle to the service station for preventive maintenance per the schedule mentioned in the service manual of the vehicle supplied by the manufacturer at a fixed time interval or distance covered by the vehicle. The service could be for an engine oil and filter change. In some instances, it may happen that the engine oil is good enough and doesn't warrant a change. For an individual, this may be a small price to pay in terms of the inconvenience of taking the vehicle to the service station for an engine oil change. However, if this was a vehicle fleet owner having thousands of vehicles, and if all the vehicles had to undergo preventive maintenance as per the OEM schedule, there could be instances where the cost of service would be very high and could have been avoided if only the vehicles whose engine oils had gone bad were replaced and serviced. Of course, the positive side of this type of maintenance is that the vehicles in the fleet are always in perfect running condition. Some automobile manufacturers are coming up with models that use a condition-based maintenance technique described in the next section.

Another example where preventive maintenance is used is for strategic reasons. In the armed forces, we would always like the weapons to fire whenever required to do so, army vehicles and naval vessels to move immediately on an order from the field commander, and so on. In such scenarios, cost is not a major criterion but functionality of the system is. Soldiers would prefer that their gun not malfunction in front of an adversary because it has not been maintained. Defense forces follow the practice of preventive or periodic maintenance for all their equipment and machines as per the schedules laid out in their service manuals. However, if all industrial machinery undergoes preventive maintenance, the overall maintenance costs could be extremely high. And to be profitable in business, one must reduce extra costs.

2.4 Predictive Maintenance

In *predictive maintenance*, maintenance is done on a machine depending on its need. The decision to conduct maintenance or not to conduct maintenance depends upon the machine's past and present condition. However, in order to know a machine's condition, additional instrumentation is required on the machine to measure its "health" parameters. This instrumentation includes transducers, signal conditioners, data acquisition units, computer-based signal analysis systems, and software-driven diagnostic routines. Preventive maintenance is a need-based maintenance that depends on the condition of the machine. This type of maintenance is also known as *condition-based maintenance*. Once the machine's condition is known, it becomes convenient, using simple mathematical models, to determine the reasons behind any impending failure of the machine or diagnose its fault condition. Once the machine's condition parameters both past and present are available in a database, again through simple mathematical regression models, the fault prognosis can be done and the remaining useful life of the machine with its present condition can be predicted.

Predictive maintenance has the additional requirement—the instrumentation system mentioned earlier—so at the time the machine is installed there is an additional expenditure for this instrumentation system. The advantage of such a maintenance technique is that the machine's condition is known to the operator at any time it is needed, which allows the operator to determine whether the machine needs to be fixed or not. Worldwide, it has been reported that over an extended time, predictive maintenance is more economical than preventive maintenance, though it is initially expensive because of the additional requirement of an instrumentation system.

The advantages of the predictive maintenance technique over the other two are many. For instance, it is economic in the long run, it provides a scope for fault prognosis, the maintenance schedule can be controlled according to the availability of resources, the spare parts inventory can be reduced, and the faults in a machine can be minimized. In the long run, this type of maintenance leads to high production rates and increased profitability. One of the major disadvantages of predictive maintenance is that it requires extra investment in the initial stage for the additional instrumentation and it needs a robust software for fault diagnosis and prognosis. Today, software with robust fault-detection algorithms is available, but for the initial runs, one must use them with the help of trained personnel so that no untested fault scenarios are misdiagnosed, leading to a disastrous consequence. Thus the benefits of predictive maintenance can be summarized as follows:

- Lower maintenance costs
- Fewer machine failures
- Less repair downtime
- Reduced small parts inventory
- Longer machine life
- Increased production
- Improved operator safety

Following are the attributes by which a predictive maintenance program can be characterized:

- User-friendly hardware and software
- Automated data acquisition

- Automated data management and trending
- Flexibility
- Reliability

Some of the techniques that are used for predictive maintenance are vibration monitoring, wear debris and oil analysis, motor current signature analysis, thermography, and nondestructive test techniques like ultrasonics, radiography, eddy current testing, and acoustic emission. These techniques will be dealt in depth in later chapters.

2.5 Enterprise Resource Planning

In any plant where there are several machines that have to be maintained, an efficient database system needs to be in place to keep the maintenance record of each machine, spare parts used, production rate, downtime if any, and so on. The above operations finally show their effect and influence on the company's balance sheet. The plant's management needs to be abreast of these figures on a continuous basis at a fixed frequency; for instance, every morning the top management of the company needs to know the number of goods produced in the last 24 hours, any serious breakdown of a machine, quantity of raw material consumed, and so on. These inputs from various operations in the plant are fed to the corresponding modules of the enterprise resource planning (ERP) software installed in a plant. So in other words, ERP helps the company management to have a bird's eye view of all operations of the plant. The machinery maintenance module is thus a very important module in the ERP system of the plant. A successful maintenance program in a plant must have the following:

- Clearly defined objectives and goals
- Measurable benefits
- Management support
- Dedicated, accountable personnel
- Efficient data collection and analysis procedure; viable database
- Communication capability
- Evaluation procedures
- Verification of new equipment condition
- Verification of repairs
- Overall profitability

2.6 Bath Tub Curve

The typical machine failure rate versus time plot is given in Figure 2.1. The plot has three distinct zones: the infant mortality zone, the useful period, and the wear-out zone. The infant mortality zone with high failure rates occurs in the early stages of the machine. There could be several reasons for such high failure rates; some of them are faulty installation at the site, ignorance and unfamiliarity of the machine operator, improper electrical power supply, nonavailability of a user or training manual, improper specifications, and choice of the machine. Once the above reasons are sorted out, the machine's failure rate reduces significantly; this state of the machine continues for a considerable time, which is known as the useful life of the machine. Finally, toward the end of the useful period, the failure rate of the machine again increases, which can be due to excessive wear and tear on the machine and fatigue failure of the machine component. Though by maintenance the failure rates can be controlled and reduced, a time comes when the cost of maintenance or upkeep is so high that it is better to completely replace the machine with a new one. The shape of the curve in Figure 2.1 is in the form of a bath tub, hence the name.

The availability of the machine is defined as the ratio of the useful period (also known as *uptime*) to the total lifespan of the machine. Maintenance engineers strive to increase the availability of a machine by decreasing the machine's downtime. The total lifespan of a machine is the summation of the uptime and downtime of the machine.

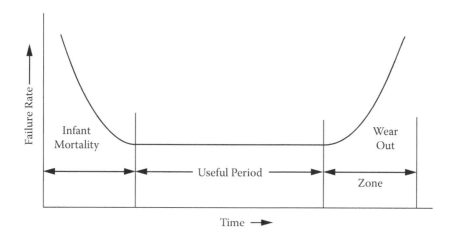

FIGURE 2.1
Bath tub curve.

2.7 Failure Modes Effects and Criticality Analysis (FMECA)

FMECA is a methodology widely used in the industry to identify and analyze all potential failure modes of the various parts of a system, determine the effects these failures may have on the system, and how to avoid the failures or mitigate the effects of the failures on the system.

FMECA can be used for the following objectives:

- Assist in selection of design alternatives with high reliability
- Ensure that all conceivable failure modes and their effects on operational success of the system have been considered
- List potential failures and identify the severity of their effects
- Develop early criteria for test planning and test equipment requirement
- Provide historical documentation for future reference
- Provide a basis for maintenance planning
- Provide a basis for quantitative reliability and availability analysis

FMECA should be initiated early in the planning process for machinery maintenance, where we are able to have the greatest impact on the equipment reliability. Following are the types of FMECAs that are prevalent in the industry:

1. *Design* FMECA is carried out to eliminate failures during equipment design, taking into account all types of failures during the whole lifespan of the equipment.
2. *Process* FMECA is focused on problems stemming from how the equipment is manufactured, maintained, or operated.
3. *System* FMECA looks for potential problems and bottlenecks in larger processes, such as entire production lines.

2.7.1 Implementation of FMECA for Machinery Maintenance

In a large plant, it would be economical or logical to have all equipment under reactive maintenance or predictive maintenance. A FMECA analysis early in the planning stage of maintenance lets one determine the critical machines that will require more attention than the rest. To implement FMECA for machinery maintenance, follow these steps:

- Define the system to be analyzed.
- Divide the system into manageable units, typically functional elements.

- Sometimes it may be beneficial to illustrate the system by a functional block diagram.
- The analysis should be carried out on as high a level in the system hierarchy as possible.
- Collect available information that describes the system to be analyzed; include drawings, specifications, schematics, component lists, interface information, functional descriptions, and so on.
- Collect information about previous and similar designs from internal and external sources, interviews with design personnel, operations and maintenance personnel, and component suppliers.

2.7.2 Risk Priority Number for FMECA

The risk related to the various failure modes in a machinery system is often presented either by a *risk matrix* or a *risk priority number* (RPN). The RPN is defined as the product of the following three numbers:

- **O:** The rank of occurrence of a failure mode
- **S:** The rank of severity of the failure mode
- **D:** The rank of ease of *not* detecting the failure

$$\text{Risk Priority Number} = O \times S \times D \qquad (2.1)$$

The ranks of O, S, and D can be anywhere from 1 to 10. So the minimum value of the RPN is 1 and the maximum value is 1000. Thus, in the maintenance planning while conducting a FMECA study on all machines in a plant, machines with a high RPN may undergo predictive maintenance and the machines with a low RPN may undergo reactive maintenance. The RPNs can vary from plant to plant or from team to team; they have no clear portability other than identifying machines within the same group on their risk assessment.

3

Fundamentals of Machinery Vibration

3.1 Introduction

Machine vibration is one of the most convenient signals for the maintenance engineer to measure and analyze and thus get to know about the machine's health condition. In this context the various basic concepts of machine vibration will be introduced in the chapter. A basic understanding of machine vibration will enable one to diagnose machinery faults and control machinery vibration.

Vibration is a representation of the motion of a body, is essentially oscillatory about a mean position, and can be periodic or aperiodic. This motion of the body can be described in any direction. However, the degree of freedom is the number of independent coordinates that can be used to describe the motions of a rigid body. For example, a rigid body in space has 6 degrees of freedom, which consist of three translatory motions and three rotational motions about these axes. Usually, for a large machinery supported at the four corners on a shop floor, the vibration motion in three directions is measured and analyzed. However, before discussing in details of the motion of bodies in more than one direction, the motion of a rigid mass in one direction is considered in order to get familiar with the associated terminology.

3.2 Single Degree-of-Freedom Motion

A rigid body of mass m is supported in the axial direction by a linear spring of stiffness k as shown in Figure 3.1. The body is only allowed to have motion in one direction. The equation of the motion of such a body is given as in Equation (3.1).

$$m\frac{d^2x}{dt^2} + kx = 0 \tag{3.1}$$

Equation (3.1) is a linear differential equation of the first order, and the response $X(t)$ of the form $x(t) = Xe^{i\omega t}$ is actually a solution to the

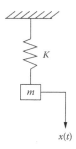

FIGURE 3.1
Single degree-of-freedom undamped system.

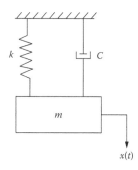

FIGURE 3.2
Single degree-of-freedom damped system.

differential equation. The constant X can be found by the initial conditions of the body. Such a response of the body is known as the *undamped free vibration response* of a harmonic oscillator. The natural frequency of such a system is given as $\omega_n = \sqrt{\dfrac{k}{m}}$.

However, no system in real life is undamped, and the oscillator can be damped as shown in Figure 3.2. The equation of motion of such a damped harmonic oscillator is given in Equation (3.2).

$$m\frac{d^2x}{dt^2} + c\frac{dx}{dt} + kx = 0 \tag{3.2}$$

The damping coefficient C is known as *viscous damping*, where the damping force is proportional to the velocity of the body. The *damping factor* is described as $\zeta = \dfrac{C}{2\sqrt{km}}$. The response to such a damped oscillator is given in Equation (3.3)

$$x(t) = A^{-\zeta\omega t} \sin(\omega_d t + \phi), \tag{3.3}$$

where, $\omega_d = \omega_n\sqrt{1-\zeta^2}$.

The systems described in Figures 3.1 and 3.2 represented linear motion. Similar systems can also represent oscillatory motions in rotation, which are otherwise known as *torsional vibrations*. The generic equation of motion representing the free vibration response of a single degree-of-freedom damped torsional vibration system is shown in Equation (3.4).

$$I\frac{d^2\theta}{dt^2} + C_t \frac{d\theta}{dt} + k_t\theta = 0 \tag{3.4}$$

where I is the rotary mass moment of inertia of the body, C_t is the torsional viscous damping co-efficient in the body, and K_t is the torsional stiffness given in N-m/rad. The rotational displacement, rotational velocity, and rotational acceleration are represented by $\theta, \frac{d\theta}{dt}, \frac{d^2\theta}{dt^2}$ respectively. In machines, and in particular gearboxes, equations of motion are used to describe the motion and thus the vibrations. However, in machines, the entire dynamic is well represented by torsional systems, and the response in the field is usually done using transverse vibration measurements. This is due to the easy availability of vibration measuring equipment for linear or transverse measurements. Though recently with the availability of laser-based torsional vibration measurement systems, machine dynamics can be better understood.

3.3 Forced Vibration Response

Equations (3.1) to (3.4) represented single degrees of freedom, where the body was not subjected to any external excitation in terms of a time-varying force or torque. However, in actual rotating machines, there are many forms of excitation, like forces due to unbalanced rotors, forces at couplings due to shaft misalignment, dynamic forces at bearing locations due to loose components, and so on. The vibration response of such machines is quite different and their study is essential for understanding the cause behind such responses. The dynamic forces or *torques* that excite the machine system can be represented by harmonic functions. In fact, any type of excitation can be mathematically represented as a sum of harmonic functions, as is done by a Fourier series expansion (more in Chapter 5). Then the principle of linear superposition can be used to determine the overall response to such types of excitations.

The equation of motion of a damped single degree-of-freedom system subjected to an external force is given by Equation (3.5)

$$m\frac{d^2x}{dt^2} + c\frac{dx}{dt} + kx = F(t), \tag{3.5}$$

where $F(t)$ is the external forcing function. For a harmonic force, the force can be represented as Equation (3.6).

$$F(t) = F_0\cos\omega_f t \qquad (3.6)$$

The excitation force is at a frequency ω_f. The system's response is very much dependent on the frequency ratio, r, which is given as $r = \dfrac{\omega_f}{\omega_n}$. The excitation frequency is sometimes known as *input frequency, excitation frequency,* or *forcing frequency.* In rotating machines, the rotational speed of the machine corresponds to this frequency.

The response of a damped harmonic oscillator to a harmonic force of a kind represented by Equation (3.6) is give by Equation (3.7).

The first expression of Equation (3.7) is known as the *transient response* and the second expression is known as the *steady-state response.*

$$x(t) = Ae^{-i\omega t}\sin(\omega_d t + \theta) + A_0\cos(\omega_f t - \phi) \qquad (3.7)$$

where $\dfrac{A_0 k}{F_o} = \dfrac{1}{\sqrt{(1-r^2)^2 + (2\zeta r)^2}}$ and $\phi = \tan^{-1}\dfrac{2\zeta r}{1-r^2}$

The normalized response of the damped harmonic oscillator is shown in Figure 3.3 (a and b).

At resonance, when $r = 1$, the forcing frequency is equal to the natural frequency of the system, and the ratio of the amplitudes decreases with an

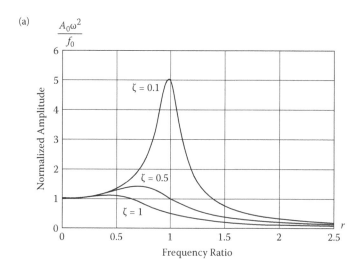

FIGURE 3.3
(a) Magnitude response of a harmonically forced damped oscillator.

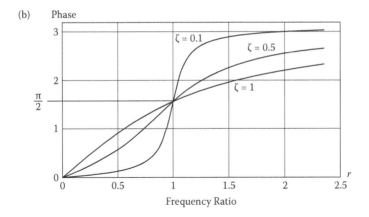

FIGURE 3.3 (*Continued*)
(b) Phase between response and force of a damped oscillator.

increase of the damping ratio. From Figure 3.3b, it can also be noticed that at resonance the phase angle between the response and the excitation for an undamped system shifts by 90°.

3.4 Base Excitation

There are instances when machine bases that are supported on some stiffness are excited by a harmonic motion, as indicated in Figure 3.4. The ratio of the steady-state response of the body X to external motion at the base Y can be represented by the expressions given in Equation (3.8) (Figure 3.5). The ratio of the forces transmitted due to the base motion is given in Equation (3.9).

$$\frac{X}{Y} = \sqrt{\frac{1+(2\zeta r^2)}{(1-r^2)^2 + (2\zeta r)^2}} \tag{3.8}$$

$$\frac{F_T}{F_y} = r^2 \sqrt{\frac{1+(2\zeta r)^2}{(1-r^2)^2 + (2\zeta r)^2}} \tag{3.9}$$

3.5 Force Transmissibility and Vibration Isolation

When a machine operates at a particular speed, or an operation like sheet-metal punching or metal forging is done at a particular rate, a periodic force is usually generated in the machine at that frequency. This force gets

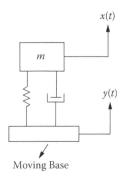

FIGURE 3.4
Body subjected to base excitation.

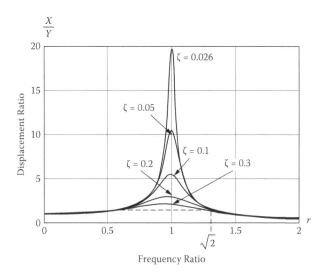

FIGURE 3.5
Ratio of the response of the body to the base excitation.

transmitted in the vertical direction to the base or foundation of the machine. The ratio of the force transmitted to the applied force is known as the *force transmissibility* and is given by Equation (3.10).

$$\frac{F_T}{F_0} = \sqrt{\frac{1+(2\zeta r)^2}{(1-r^2)^2+(2\zeta r)^2}} \qquad (3.10)$$

Notice in Figure 3.6 that at a value of $r > \sqrt{2}$, this ratio is less than unity. In practice, the stiffness and damping present between the machine and the foundation play an important role in the amount of force that gets transmitted.

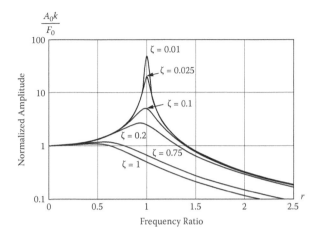

FIGURE 3.6
Force transmissibility ratio.

Vibration isolators are available on the market that are selected based on two important parameters, one of which is the maximum deflection that allows the base and the damping to support the payload, as given by Equation (3.11). Depending on the configuration of the isolator arrangement (usually for parallel arrangements of the isolators at the base), the effective spring stiffness is given by Equation (3.12).

$$K_{eff}\delta_{max} = W \tag{3.11}$$

$$K_{eff} = Nk_{spring} \tag{3.12}$$

For vibration and force isolation, usually helical springs, wire rope springs, and elastomeric pads are used in industry. A typical wire rope spring used as a machinery mount for vibration isolation is shown in Figure 3.7. These types of springs have some inherent damping present in them, which can be anywhere from 5% to 15%. The second important parameter in the selection of isolators is that of stiffness, so that the effective natural frequency of the system in the direction of the force transmission is less than $\dfrac{\omega_f}{\sqrt{2}}$.

3.6 Tuned Vibration Absorber

The vibration response of a system is predominantly at the forcing frequency. However, there are instances where the forcing frequency may be equal to the natural frequency of the system; then the system will undergo resonance and undergo large motions. A designer usually ensures that the

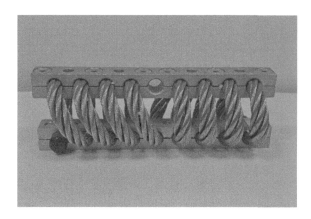

FIGURE 3.7
Wire rope vibration isolator.

operating frequency or the forcing frequency of excitation of the machine is not close to its natural frequency. However, in instances where such a coincidence of the frequencies cannot be avoided, another secondary spring mass system is usually attached to the primary machine, so that the natural frequency of the primary system is shifted and the condition of resonance is avoided. Such a secondary system is known as a *tuned dynamic absorber*. A tuned absorber attached to a primary mass is shown in Figure 3.8a. The secondary system is selected so that the natural frequency of both the primary and secondary systems are the same, as shown in Equation (3.13).

$$\omega_n = \sqrt{\frac{k_p}{m_p}} = \sqrt{\frac{k_s}{m_s}} \tag{3.13}$$

Such tuned absorbers are used in many industrial applications; for example, to arrest the vibrations of driveline shafts in automobiles, tuned vibration absorbers are attached at the bearing supports; and tuned absorbers attached to high-voltage transmission cables reduce their vibrations from wind force excitations. A dumb bell–shaped tuned absorber known as the Moose damper is being evaluated in the laboratory on a slip table at the Indian Institute of Technology Kharagpur for its natural frequencies, and is shown in Figure 3.8b.

3.7 Unbalanced Response

A common machinery defect is *rotating unbalance*. If a rotating machine of mass M has an unbalance mass, m_0, at a radius of e from the center of rotation, and the machine is only allowed to have vertical motion as shown

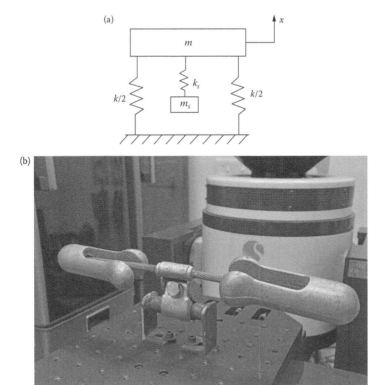

FIGURE 3.8
(a) Tuned absorber, (b) Moose damper under resonance test on slip table.

in Figure 3.9, the amplitude of vibration response of the unbalanced machine is given by Equation (3.14). The response of such a unbalanced system is shown in Figure 3.10.

$$X = \frac{m_0 e}{M} = \frac{r^2}{\sqrt{(1-r^2)^2 + (2\zeta r)^2}} \tag{3.14}$$

In order to reduce this response at operating speeds, care is taken to reduce the unbalance mass, m.

3.8 Characteristics of Vibrating Systems

In machinery condition monitoring, the response of a system is measured, analyzed, and an understanding of the system is achieved. However, the vibrating system's characteristics can be known by measuring its transfer

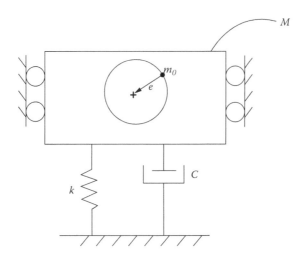

FIGURE 3.9
Unbalanced mass in a rotating machine.

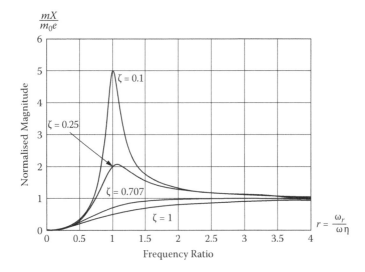

FIGURE 3.10
Unbalanced response.

function or frequency response function. The transfer function is usually represented in the Laplace domain and the frequency response function (FRF) is represented in the frequency domain. The FRF is the ratio of the response of the system to its excitation. There is a phase difference between the response and the excitation and FRF is a complex quantity.

TABLE 3.1

Forms of Frequency Response Function (FRF) for
Forced Type of Excitation

Response	FRF (Response/ Excitation)	Inverse of FRF
Displacement	Compliance	Stiffness
Velocity	Mobility	Impedance
Acceleration	Inertance	Apparent mass

Generally, three forms of FRF are prevalent depending upon the type of vibration response. The common representations of the FRF are shown in Table 3.1. For a single degree-of-freedom damped harmonic oscillator, the impedance is given by Equation (3.15).

$$Z(\omega) = C + j(m\omega - k/\omega) \tag{3.15}$$

From Equation (3.15) it is seen that at resonance, the impedance is purely dependent on the damping coefficient, C. Thus, the vibration amplitudes at resonance can be decreased by introducing additional damping at appropriate places on the machine or structure.

3.9 Vibration of Continuous Systems

In previous sections, the vibrations of a single degree-of-freedom system consisting of a rigid mass were considered. However, any mechanical structure can be thought of as being made of many such small rigid masses, each having a degree of freedom. Thus, the entire mechanical structure can theoretically have infinite degrees of freedom. To describe the vibrations of such continuous systems, the equations of motion to determine the displacement response of such systems to forces can be determined by the solutions of the equations of motion. A list of such equations of motion for free vibration dynamic response of continuous systems is provided in Table 3.2. This table provides the equations of motion of some commonly occurring engineering structures like bars, rods, beams, and plates. The natural frequencies of these structures can be estimated once their stiffness and mass are known. If these structures are excited by a dynamic force, the response can be estimated by solving the differential equations. There are many methods of solving these equations for an

TABLE 3.2

Equation of Motion for Few Continuous Systems

Name of System	Descriptive Figure	Equation of Motion	Important Parameters
Longitudinal bar		$\dfrac{\partial^2 u}{\partial t^2} - c^2 \dfrac{\partial^2 u}{\partial x^2} = 0$	$c^2 = \dfrac{E}{\rho}$
Torsional bar		$\dfrac{\partial^2 \theta}{\partial t^2} - c^2 \dfrac{\partial^2 \theta}{\partial x^2} = 0$	$c^2 = \dfrac{G}{\rho}$
Longitudinal beam (Euler-Bernoulli)		$\rho A \dfrac{\partial^2 w}{\partial t^2} - EI \dfrac{\partial^4 \theta}{\partial x^4} = 0$	
Longitudinal beam (Timoshenko beam)		$\dfrac{\rho I}{GA_s} \dfrac{\partial^4 w}{\partial t^4} + \dfrac{\partial^2 w}{\partial t^2} - \left(\dfrac{1}{A} + \dfrac{EI}{GA_s} \right) \dfrac{\partial^4 w}{\partial t^2 \partial x^2} + \dfrac{EI}{\rho A} \dfrac{\partial^4 w}{\partial x^4} = 0$	$A_s = \dfrac{A}{k}$
Rectangular plate (Kirchoff's theory)		$\rho h \dfrac{\partial^2 w}{\partial t^2} + D \left(\dfrac{\partial^4 w}{\partial x^4} + 2 \dfrac{\partial^4 w}{\partial x^2 \partial y^2} + \dfrac{\partial^4 w}{\partial y^4} \right) = 0$	$D = \dfrac{Eh^3}{12(1-v^2)}$

analytical solution using the principles of energy conservation. When the structures become large and complex, numerical methods like the finite element method are used.

3.10 Mode Shapes and Operational Deflection Shapes

Mode shapes are the loci of points on the structure that are displaced when the structure is vibrating at one of its natural frequencies. On the structure, there are points where there is no displacement at a particular natural frequency and points at which there is maximum displacement; these are known as the *antinode points*. Figure 3.11 shows the first three mode shapes of a cantilever beam at its three natural frequencies.

During vibration measurement, it is usually advisable not to mount the transducer at a node location, since the vibration at that particular natural frequency at the node point is zero. By experimental modal analysis, the mode shapes of the structure at its natural frequencies can be measured. Many commercial postprocessing software programs are available where the measured vibration data at all positions around the structure can be animated to give a view of the relative motion of the different positions on the structure when they vibrate at one of their natural frequencies.

Unlike mode shapes, operational deflection shapes (ODSs) of structures are the relative vibrations of all points on the structure at any

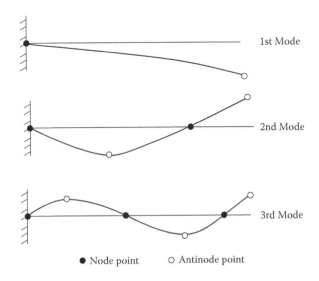

FIGURE 3.11
First three mode shapes of a cantilever beam under transverse vibration.

given frequency. The measurement of ODSs helps in understanding if two parts of a body are touching each other or the amount of clearance between them. The ODS is usually measured at the operating frequency of the machine. The values of the displacement as measured by the ODS are dependent on the actual operating condition of the machine and are thus dependent on the loads in terms of magnitude, direction, and location on the machine.

3.11 Experimental Modal Analysis

In many instances, the vibration at the measurement points in machines may increase due to an unavoidable resonance condition. This resonance happens because the external forcing frequency has matched that of one of the resonant frequencies of the machine. The source of the external forcing frequency is usually due to the operating speed of the machinery, and many times it is observed that by shifting the operating speed, the resonant condition or the maximum vibration suddenly decreases. Of course, the machines are designed so that the system's natural or resonance frequencies are not close to the machine's operating speed. However, in many processing plants for example, if there is a stirrer mounted on top of a storage tank, it has been observed that sudden violent vibrations occur on the tank at a certain filling position. This is due to the fact that, at a certain minimum level of the tank, the mass of the storage tank changes, which increases the corresponding natural frequency. While troubleshooting such excessive vibration phenomena, it is desirable to do an in situ estimation of the natural frequency of the system.

Experimental modal analysis is an experimental technique to determine the natural frequencies, associated damping at the natural frequencies, and the mode shapes of the few major modes of the machinery. To determine the frequency response of the system, the system has to be excited and the response measured at the desired points. Depending upon the number of excitation points and the response points, the system can be called as a single-input single-output (SISO) or a multiple-input multiple-output (MIMO) system. Depending on the response and excitation type, the FRFs are obtained as described earlier in Table 3.1. The two most common types of excitation system used for modal analysis are the impact type, using an instrumented hammer with a force transducer at the tip of the hammer, and the random excitation type, using an electromagnetic exciter driven by a random noise. At the tip of the stinger that is attached to the electromagnetic exciter a force transducer is attached. The response of the structure is usually measured by any contact or noncontact vibration transducer. Usually piezoelectric accelerometers are used for the response measurements, although care is to be taken that the weight of the accelerometer does not load the device under test (DUT). For light structures, laser vibrometers are preferred. The peaks in the FRF correspond to the natural frequencies of the system. If the response measurements are acquired and overlaid on the geometrical drawing of the DUT, along with the relative phase measurements with a reference transducer that is fixed on the body, the relative measurements at all points around the body at the resonant frequency will give the mode shape of the body at that frequency. The points where the vibration of the body are zero or close to a minimum are known as the *node points* and the points on the body where the modes are at maximum are known as the *antinodes*, though as mentioned earlier, similar plots at any other frequencies while the machine is operating at its rated speed are known as *operational deflection shapes*. Figure 3.12 shows a schematic diagram of experimental modal analysis on a test structure using a random excitation through an electromagnetic shaker.

From the resonant peaks at the natural frequencies, the modal damping can be estimated by measuring the Q factor of the system, which is given in Equation (3.16). One such resonant peak is shown in Figure 3.13, which is used to estimate the damping at the resonant or modal frequency. In Equation (3.16), A is the amplitude of the FRF at the natural frequency of f_0, and f_1 and f_2 are frequencies on either side of the peak where the amplitude of the FRF is 0.707A.

$$Q = \frac{1}{2\zeta} = \frac{f_0}{f_2 - f_1} \tag{3.16}$$

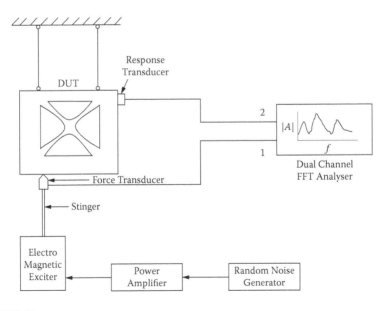

FIGURE 3.12
Schematic for experimental modal analysis.

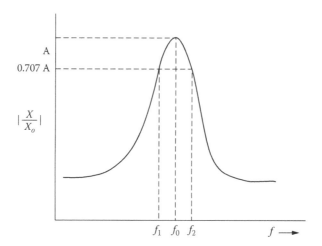

FIGURE 3.13
Resonant peak for damping estimate by half-power method.

In the field, while doing a quick estimate to determine the natural frequencies, a *bang* test is done, where only the frequency analysis of the response is done at the instant the structure is excited by a bang. The natural frequencies are identified by noticing the peaks in the frequency plot of the response; this is true because the impulse provided by the bang excites the DUT at all frequencies. The system's response at its natural frequencies will be high due to resonance as is indicated by the peaks in the response spectrum.

4

Rotordynamics

4.1 Introduction

A very important component in any machinery is a shaft that rotates and is supported at two ends by bearings. This shaft may have a disc mounted on it, which could be like a gear, pulley, and so on. The dynamics of the shaft at different rotating speeds is very crucial to designers, because the loads at the bearings may vary, and oscillations in the shaft may occur which may lead to conditions of instability. Conditions of resonance may occur at certain operating speeds of the shaft. These issues will be briefly addressed in this chapter.

4.2 Simple Rigid Rotor-Disc System

A simple rotor-disc system is shown in Figure 4.1. The disc is centrally located on the shaft. The load of the disc is supported at the bearings at the two ends. This simple rotor-disc system is known as a *Jeffcott rotor*. In heavy machineries, as the weight of the central disc increases, the load on the bearings also increases. Thus it becomes a challenge to support heavy loads, because the antifriction bearings can only support loads to their designed limits. The bearings in such large machines are fluid filled and are known as *journal bearings*, which will be discussed later. One such large system is a turbo-generator in a power plant, a schematic of which is shown in Figure 4.2.

4.3 Unbalance Response and Critical Speed

For the disc shown in Figure 4.3, let the center of mass be at a location G instead of the center of the shaft at E. The eccentricity EG is denoted as a. When the shaft rotates at a frequency ω, there will be transverse vibration in the directions z and y as shown in the figure. Due to this unbalance, the

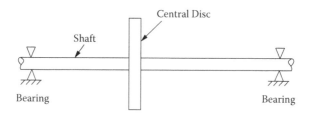

FIGURE 4.1
Jeffcott rotor.

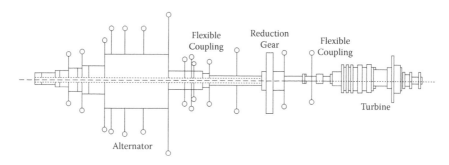

FIGURE 4.2
Schematic of a large turbo-generator system.

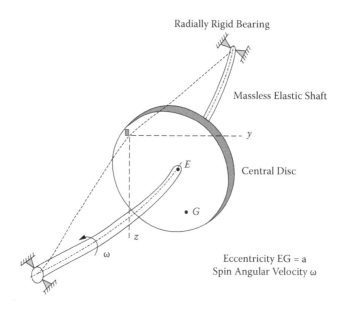

FIGURE 4.3
An unbalanced disc on a elastic shaft.

shaft will bow and spin about the axis; this phenomenon is known as the *whirling of shafts*. Whirling is the angular motion of the deflected shaft rotating about the neutral axis of the shaft. In large turbo machines, this would be of concern, since if the amplitudes of bow are large, there may be a failure of the system, due to fatigue and associated loadings. In the design of the operating speeds of the rotor system, such speeds are to be avoided. The response of the shaft in the y and z directions are given by Equations (4.1) and (4.2).

$$y(t) = \frac{ar^2}{\sqrt{(1-r^2)^2 + (2\xi r)^2}} \sin\left(\omega t - \tan^{-1}\frac{2\xi r}{1-r^2}\right) \tag{4.1}$$

$$z(t) = \frac{ar^2}{\sqrt{(1-r^2)^2 + (2\xi r)^2}} \sin\left(\omega t - \tan^{-1}\frac{2\xi r}{1-r^2}\right) \tag{4.2}$$

The amplitude of the radial response of the shaft is given in Equation (4.3). Here, X is the magnitude of $z(t)$. This indicates that the distance between the shaft and its neutral axis is constant and has a magnitude given by Equation (4.4).

$$|r(t)| = \sqrt{z^2 + y^2} = X\sqrt{\sin^2(\omega t - \phi) + \cos^2(\omega t - \phi)} = X \tag{4.3}$$

$$X = \frac{ar^2}{\sqrt{(1-r^2)^2 + (2\xi r)^2}} \tag{4.4}$$

$$\phi = \tan^{-1}\frac{2\xi r}{1-r^2} \tag{4.5}$$

For the case of $m = m_0$, the normalized amplitude as a function of the frequency ratio, r, is that given in Figure 3.10. At resonance the when $r = 1$, for a lightly damped system, the response has high amplitudes. The rotor's speed in this instance is known as its critical speed. In turbo-machineries like that shown in Figure 4.2, there will be numerous critical speeds and a designer has to decide on safe frequency zones of operation. Operating speeds creating resonances are clearly shown in a Campbell diagram, shown in Figure 4.4.

In the case of synchronous whirl, the whirling frequency is the same as the (spin frequency) angular speed ω at which the disk rotates about the shaft.In many instances the spin frequency is the operating speed of the shaft.

4.4 Journal Bearings

In rotating machineries, the bearings have an important function of supporting the shafts and the associated loads. At the supports, they must also provide much less friction. Less friction is one of the reasons for

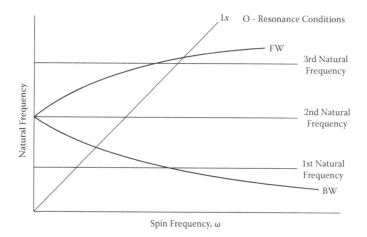

FIGURE 4.4
Campbell diagram for multidisc rotor.

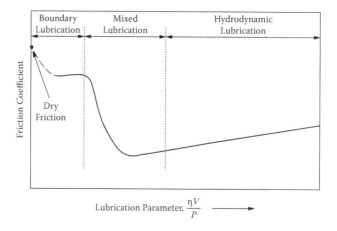

FIGURE 4.5
Regimes of lubrication.

using a fluid film bearing, wherein hydrodynamic lubrication is present. This fluid friction is dependent on the fluid viscosity, speed, and load as shown in Figure 4.5.

A journal bearing is shown in Figure 4.6, where the journal is eccentric in the sleeve by an amount e. The region between the journal and the sleeve is filled with a lubricating oil of a certain viscosity. Due to the converging–diverging wedge formed between the journal and the shaft, a pressure is built up toward the center of the sleeve at high rotational speeds. This fluid pressure is responsible for supporting the load W on the shaft. When the shaft is not rotating, the journal sits on the sleeve; as a result, the sleeve

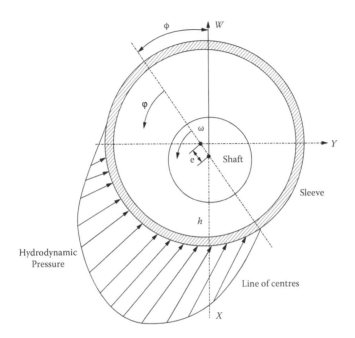

FIGURE 4.6
Journal bearing.

material is subjected to wear during repeated starts and stops. The sleeves are actually lined with soft bearing material known as Zn/Sn alloys, and can be replaced when they wear out. The fluid film thickness is an important aspect in journal bearings and the position of the journal in the bearings is usually measured at two mutually perpendicular directions and monitored.

When the loads are very high in precision shafts rotating at high speeds, and the pressure built up in journal bearings is not adequate, externally pressurized fluid is provided to the journal. Such bearings are known as hydrostatic bearings. In hydrostatic bearings, since the fluid pressure is always present, the journal never comes in contact with the sleeve.

Due to damping provided by the fluid in the journal bearing, the vibration response of the rotating shaft at any speed, in particular at resonance, can be controlled. This prevents any dynamic instability in the rotor-bearing system.

4.5 Oil Whirl and Oil Whip

In the case of journal bearings when the shaft rotates at a speed ω, the oil in the wedge can be thought to have a speed ω next to the journal and zero next to the stationary sleeve. Thus, the oil whirls at a speed about 0.42 to 0.48 ω, which is a characteristic frequency that is present in the vibrations from rotors supported by journal bearings.

Oil whip occurs when the whirling frequency coincides with one of the natural frequencies of the rotor system.

4.6 Squeeze Film Dampers

In rotor-bearing systems supported on journal bearings, the amplitude of the vibration response of shafting systems may become large, and at certain operating speeds may lead to rotor instability and catastrophic failure. In journal bearings, squeeze film dampers consisting of external pads are used which, due to relative motion between them, provide additional damping by a squeezing motion.

4.7 Condition Monitoring in Large Rotor Systems

Considering the above aspects of rotor-bearing systems, the following are usually done to monitor rotor systems:

1. Journal bearing clearance and film thickness monitoring
2. Shaft relative displacement in the horizontal and vertical planes
3. Journal bearing oil quality and wear debris monitoring
4. Vibration monitoring to detect oil whirl frequency

5

Digital Signal Processing

5.1 Introduction

In machinery condition monitoring, a decision on the condition of a machine is dependent on the nature of its signals. For example, if the machine produces excessive noise and vibration, one gets a clue that something is not normal with the machine. These signals are usually time varying and need to be understood because they carry the information from the machine to the decision system, which can be a developed software or a person. Here we will focus our attention on a very fundamental signal, given by Equation (5.1).

$$x(t) = A\sin(\omega t + \phi) \tag{5.1}$$

This signal is graphically represented in Figure 5.1. Certain information conveyed by the signal in Figure 5.1 can be understood in terms of its duration, time period, amplitude, and so on. However, there are many other features of the signal that are not obvious, but can be computed with the expressions in Table 5.1. In Table 5.1, x_i is the value of the signal at any time instance. These features bring out the differences between the signals, but real-world measured signals, which are the net result of many events occurring at the same time, look like the one shown in Figure 5.2. A signal analyst needs to determine the features of such signals so that a proper decision can be made on the source of this signal, which in our case is the machine whose health condition needs to be determined.

5.2 Classification of Signals

Signals essentially convey information, features of which can be constant over a period of time or vary with time. Based on the above condition, signals can be classified as stationary or nonstationary. The signals whose statistical features as computed using Table 5.1 do not change with time and are known as *stationary signals*. Stationary deterministic signals have specific

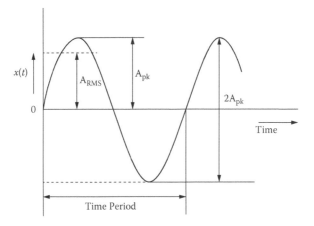

FIGURE 5.1
A sine wave signal.

distinct frequency components. Signals from rotating machines operating at a constant rotational speed are examples of stationary deterministic signals. Random deterministic signals are well characterized by their statistical features, such as mean, standard deviation, variance, and so forth. Many real-world signals are random, for example, noise produced by rain drops on a roof. Unlike deterministic signals, the spectral content of random signals is continuous. The time domain statistical features of nonstationary signals change with time. The most common nonstationary signal is a transient type that is due to an impact like the sound of something falling from a height or a slamming door. In some situations, these signals can be of the continuous repetitive type, like the noise produced by a jackhammer.

5.3 Signal Analysis

To obtain meaningful information from signals, they need to be analyzed. Analysis of stationary signals can be done in the time domain or in the frequency domain. The time domain features give an overall impression of the time domain signal. In many machines, the frequency of occurrence of a mechanical event is related to the dynamics of the process; for instance, in a ceiling fan with three blades that is used to circulate air in a room, in every shaft rotation of the fan, the air at a particular location is "chopped" three times. In other words, the fan blade pass frequency is three times the rotational (speed) of the fan. Thus, if a microphone was placed near the ceiling fan to measure the fan noise, one would notice a high amplitude of the measured noise signal, in particular at the fan blade pass frequency. Thus, every mechanical event or component has a characteristic frequency

TABLE 5.1

Time Domain Features of Signals

Mean	$$\bar{x} = \frac{1}{N}\sum_{i=1}^{N} x_i$$
Max	$x_{max} = \max |x_i|$
Min	$x_{min} = \min |x_i|$
Range	$x_{range} = x_{max} - x_{min}$
Sum	$$x_{sum} = \sum_{i=1}^{N} x_i$$
RMS	$$x_{rms} = \sqrt{\frac{1}{N}\sum_{i=1}^{N} x_i^2}$$
Standard Deviation	$$\sigma = \sqrt{\frac{1}{N}\sum_{i=1}^{N}\left(x_i - \bar{x}\right)^2}$$
Variance	$$\sigma^2 = \frac{1}{N}\sum_{i=1}^{N}\left(x_i - \bar{x}\right)^2$$
Kurtosis	$$x_{kur} = \frac{\sum_{i=1}^{N}\left(x_i - \bar{x}\right)^4}{(N-1)\sigma^4}$$
Skewness	$$x_{ske} = \frac{\sum_{i=1}^{N}\left(x_i - \bar{x}\right)^3}{(N-1)\sigma^3}$$
Crest factor	$$c_f = \frac{x_{max}}{x_{rms}}$$
Form factor	$$s_f = \frac{x_{rms}}{\bar{x}}$$

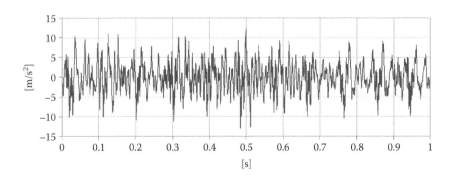

FIGURE 5.2
A typical vibration signal from a reciprocating compressor.

in the signal produced by the system. So, by performing a frequency domain analysis, one can capture the predominant frequency or frequencies where mechanical events are occurring. Every mechanical component or process in a machine, while in operation, will have a distinct frequency of occurrence in the signal being produced by it. In other words, a frequency domain analysis of a signal provides a mechanical signature of the system. If there is a change in the dynamics because of a fault or a defect in the system, then there could be a corresponding change in the amplitude of the signal at that particular characteristic frequency. In fact, this is the reason why signal processing is almost a prerequisite for anyone involved in machinery condition monitoring or fault diagnostics. Of course, for frequency domain signal analysis we have assumed that the signal is stationary. However, for signals that occur only for a small time instant, like an impact, an analysis in the joint time–frequency domain is usually done. There are many such analysis methods available, like the short-time Fourier transform (STFT) and the wavelet analysis, details of which can be found in any advanced book on signal processing.

Some common dynamic signals encountered in various engineering applications are shown in Figure 5.3.

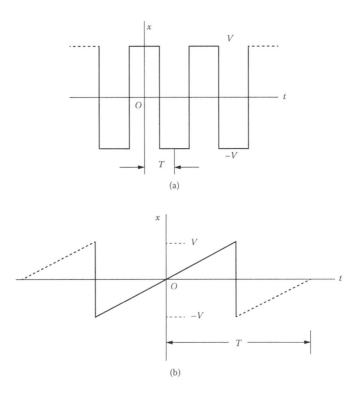

FIGURE 5.3
Some common time varying signals; (a) square wave, (b) sawtoothed wave.

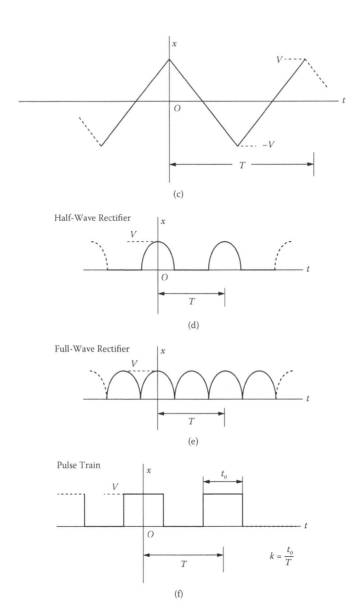

FIGURE 5.3 (Continued)
Some common time varying signals; (c) triangular wave, (d) half wave rectified wave, (e) full wave rectified wave, and (f) pulse train.

Signal beating is a phenomenon that occurs when two signals that are very close in frequency are present, and the resultant signal is given as a summation of such independent signals. This phenomenon is encountered particularly on shop floors when two machines are operating at close to the same speeds; one usually hears a successive waning and waxing of sound.

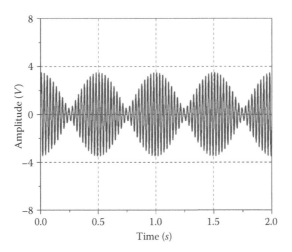

FIGURE 5.4
Time domain representation of beating signal.

Figure 5.4 shows a signal with beats. The beat frequency is the difference between the frequencies of two signals with frequencies very close to one another.

5.4 Frequency Domain Signal Analysis

5.4.1 Fourier Series

Jacques Fourier (1768–1830) was a French mathematician who showed that periodic functions can be represented by a series of sinusoids in the form of Equation (5.2).

$$x(t) = \frac{a_0}{2} + \sum_{n=1}^{\infty} \left[a_n \cos(n\omega_0 t) + b_n \sin(n\omega_0 t) \right] \tag{5.2}$$

where

$$a_n = \frac{2}{T_0} \int_{t}^{t+T_0} x(t)\cos(n\omega_0 t)dt \quad \text{for} \quad n = 0,1,2,\dots \tag{5.3}$$

$$b_n = \frac{2}{T_0} \int_{t}^{t+T_0} x(t)\sin(n\omega_0 t)dt \quad \text{for} \quad n = 1,2,3,\dots \tag{5.4}$$

$$a_0 = \frac{2}{T_0} \int_t^{t+T_0} x(t)\,dt \tag{5.5}$$

and the fundamental angular frequency of a periodic time history is $\omega_0 = 2\pi/T_0$. Once a_n, b_n and a_0 are estimated using Equations (5.3) to (5.5) the periodic function can be represented in the frequency domain as given in Equation (5.2). The Fourier series expansion for some standard mathematical functions are given in Table 5.2. The amplitude of the function at its nth harmonic can be computed using Equation (5.6).

$$A_n = \sqrt{a_n^2 + b_n^2} \tag{5.6}$$

The Fourier series of a function can also be represented in the complex domain and it can be represented in either the real-imaginary format or magnitude-phase format.

5.4.2 Fourier Integral

However, the real-world signals we come across in machinery conditions are not periodic at all times. The Fourier series coefficients of two pulses separated by a time $T_0 = \beta T$ as shown in Figure 5.5 are given in Equations (5.7) to (5.9). It would be interesting to investigate when T_0 becomes very large.

TABLE 5.2

Fourier Series of Waves Shown in Figure 5.3

Wave Shape	Fourier Series
Square Wave	$x(t) = \dfrac{4V}{\pi}\left(\cos\omega_0 t - \dfrac{1}{3}\cos 3\omega_0 t + \dfrac{1}{5}\cos 5\omega_0 t - \dfrac{1}{7}\cos 7\omega_0 t + \ldots\right)$
Triangular Wave	$x(t) = \dfrac{8V}{\pi^2}\left(\cos\omega_0 t + \dfrac{1}{9}\cos 3\omega_0 t + \dfrac{1}{25}\cos 5\omega_0 t + \ldots\right)$
Sawtoothed Wave	$x(t) = \dfrac{2V}{\pi}\left(\sin\omega_0 t - \dfrac{1}{2}\sin 2\omega_0 t + \dfrac{1}{3}\sin 3\omega_0 t - \dfrac{1}{4}\sin 4\omega_0 t + \ldots\right)$
Half-Wave Rectifier	$x(t) = \dfrac{V}{\pi}\Big(1 + \dfrac{\pi}{2}\cos\omega_0 t + \dfrac{2}{3}\cos 2\omega_0 t - \dfrac{2}{15}\cos 4\omega_0 t + \dfrac{2}{35}\cos 6\omega_0 t$ $- \ldots (-1)^{n/2-1}\dfrac{2}{n^2-1}\cos n\omega_0 t\ldots\Big)\quad n\quad\text{even}$
Full-Wave Rectifier	$x(t) = \dfrac{2V}{\pi}\Big(1 + \dfrac{2}{3}\cos 2\omega_0 t - \dfrac{2}{15}\cos 4\omega_0 t + \dfrac{2}{35}\cos 6\omega_0 t$ $- \ldots (-1)^{n/2+1}\dfrac{2}{n^2-1}\cos n\omega_0 t\ldots\Big)\quad n\quad\text{even}$
Pulse Train	$x(t) = V\left[k + \dfrac{2}{\pi}\left(\sin k\pi \cos\omega_0 t + \dfrac{1}{2}\sin 2k\pi \cos 2\omega_0 t + \ldots + \dfrac{1}{n}\sin nk\pi \cos n\omega_0 t\right)\right]$ $k = t_0/T$

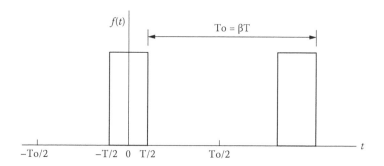

FIGURE 5.5
Two pulses separated in time.

$$a_n = \frac{2A}{n\pi}\sin\left(\frac{n\pi}{\beta}\right) \qquad n = 1,2,3,\dots \qquad (5.7)$$

$$b_n = 0 \qquad\qquad n = 1,2,3,\dots \qquad (5.8)$$

$$a_0 = \frac{2A}{\beta} \qquad\qquad (5.9)$$

It is observed from Equation (5.7) to (5.9) that the amplitudes of the Fourier series coefficients decrease as the time interval T_0 becomes large in comparison to the pulse duration T. For large values of T_0 tending to infinity, defining a new Fourier coefficient as $A(\omega) = a_n T_0$, and substituting for T_0 tending to infinity, $\omega_0 = \dfrac{2\pi}{T_0} \equiv d\omega$ and $n\omega_0 \equiv \omega$, we have

$$A(\omega) = \underset{T_0 \to \infty}{Lt}\ a_n T_0 = 2\int_{-T/2}^{T/2} f(t)\cos(\omega t)\,dt \qquad (5.10)$$

For $T \to \infty$, we have

$$A(\omega) = 2\int_{-\infty}^{\infty} f(t)\cos(\omega t)\,dt \qquad (5.11)$$

and similarly,

$$B(\omega) = 2\int_{-\infty}^{\infty} f(t)\sin(\omega t)\,dt \qquad (5.12)$$

Thus, by substitution of the Fourier coefficients in Equation (5.2), the Fourier integral expression of any function that is not periodic can be obtained as

$$f(t) = \frac{1}{2\pi} \int_0^\infty \left[A(\omega)\cos(\omega t) + B(\omega)\sin(\omega t) \right] d\omega, \tag{5.13}$$

the amplitude of the Fourier coefficients of the function, $f(t)$, at a frequency of ω are given as

$$F(\omega) = \sqrt{A^2(\omega) + B^2(\omega)}. \tag{5.14}$$

5.4.3 Discrete Fourier Transform

In the evaluation of the Fourier integral coefficients in Equations (5.11) and (5.12), it is seen that a mathematical expression of the function $f(t)$ is required. Signals that are acquired from a machine for the purpose of condition monitoring are of different frequencies and characteristics that cannot be described by a single mathematical expression, but can perhaps be reconstructed in the time domain, if the data are somehow obtained at very closely spaced time intervals. This is quite similar to drawing a graph in the X-Y plane if for all closely spaced abscissa points, the corresponding ordinates are available. The aspects regarding data acquisition are discussed later. The Fourier integral can be numerically estimated and the Fourier coefficients at discrete frequency intervals of Δf obtained as shown in Equation (5.15).

$$X(k\Delta f) = \Delta t \sum_{n=0}^{N-1} x(n\Delta t)e^{-j2\pi k\Delta f n\Delta t} \tag{5.15}$$

where k is an index.

From Equation (5.15) it is seen that to compute the Fourier transform of a signal of N discrete data points, N^2 number of complex mathematical operations are required. The discrete Fourier transform (DFT), when numerically implemented in the early days when computing resources were scarce, was a computational challenge. In the early 1960s, a new computation algorithm was developed where $N \log_{10} N$ computational operations were required instead of the N^2 operations. Using this new algorithm, there was substantial savings in the computation time needed to convert a measured real signal to its frequency domain with its complex Fourier coefficients. The $N \log_{10} N$ algorithm required that the total number of data points N, required for computation, should be a power of 2. That is, the total number of data points should be 128, 256, 512, 1024, 2048, 4096, and so on. This algorithm was a runaway success and came to be known as the fast Fourier transform (FFT). With present-day computers and software-based systems being of high computation speed, one hardly notices the difference in the computation time between computing DFT or FFT.

5.5 Fundamentals of Fast Fourier Transform

Any real analog signal $x(t)$, as shown in Figure 5.6, that has been digitized as x_i and acquired in an array of size $[1 \times N]$, can be transformed into the complex frequency domain using the transformation Equation (5.15). The Fourier coefficients of the signal are given for every $X(k\Delta f)$, which are complex. For a total of N data points, the Fourier coefficients in the positive frequency axis are present for $N/2$ points. There are few fundamental relations between the time domain sampling interval, Δt; sampling frequency, fs; total number of data points, N; total time length, T; frequency resolution f; and the maximum frequency, F_{max} of the Fourier spectrum, as given by Equations (5.16) to (5.19) and illustrated in Figure 5.7.

$$f_s = \frac{1}{\Delta t} \tag{5.16}$$

$$\Delta f = \frac{1}{T} = \frac{1}{N\Delta t} \tag{5.17}$$

$$F_{max} = \frac{N}{2} \Delta f \tag{5.18}$$

When signal analysis is performed, the user usually has a choice to select N (the number of data points), and depending on the sampling frequency, the desired frequency resolution is obtained. To reduce the low-pass filtering

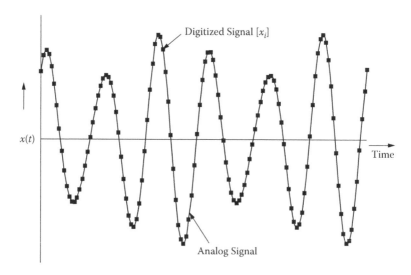

FIGURE 5.6
Digitized data.

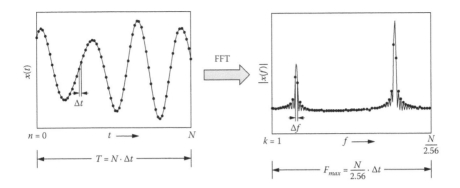

FIGURE 5.7
Illustration of time-to-frequency domain transformation.

effects at high frequencies in many commercial signal analyzers, while performing FFT, it is preferred that the number of points be reduced by 2.56 rather than 2, as given in Equation (5.19).

$$F_{\text{max}} = \frac{N}{2.56} \Delta f \qquad (5.19)$$

So, for example, when 1024 time domain data points are considered, only 400 data points are used to determine the maximum frequency in the frequency spectrum of the signal. In commercial FFT analyzers, the total number of data points whose FFT results are displayed is known as *lines of FFT*.

Few decades ago, fast Fourier transform algorithms were implemented in commercial signal analyzers, which were hardware-based devices with digital electronics. Today, user-friendly graphical user interface (GUI) software is common across all computer platforms, along with the necessary hardware, which will both acquire the time data and analyze the signal by FFT algorithm, among many other important functions like data display and storage. However, users of these FFT analyzers must consider few important aspects of these types of FFT analyzers.

Frequency resolution must be fine enough so that small differences in the frequency of a signal can be detected. This can be achieved by increasing data size. If the frequency resolution is not fine enough, small changes in the frequency of a signal can go undetected, and this is known as the *picket fence effect*.

Another important aspect is the correct amplitude estimation of the signal. When FFT is computed in one block of data, it is assumed that the data in the next block is identical; however, since the data are taken in blocks of N data points, there may be instances where an artificial discontinuity of the signal is created, as shown in Figure 5.8. Such discontinuity at the ends of the data block create energy leakage. Errors in amplitude estimation during FFT are due to this leakage error. The leakage error is reduced by multiplying

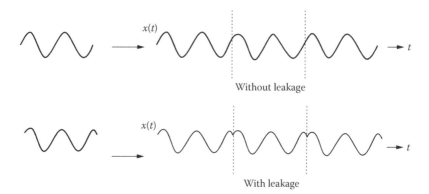

$x(t)$ Without leakage

$x(t)$ With leakage

FIGURE 5.8
Discontinuity in signal responsible for leakage error.

TABLE 5.3

Common Window Functions

Window Type	Signal
Rectangular	Impulse excitation
Exponential	Impulse response
Hanning	Periodic signal for best frequency estimation
Flattop	Periodic signal for best amplitude estimation

the sampled data with a window function, so that ends have a value of zero. Some of the common windowing function used for FFTs of different types of signals are given in Table 5.3.

5.5.1 Representation of Fourier Transform Data

The Fourier transform of a real signal $x(t)$ is a complex number $X(f) = X(k\Delta f)$; the complex number can be represented in the real/imaginary format or the magnitude phase format as given in Equation (5.20) and (5.21)

$$X(f) = X_{real} + iX_{imag} \tag{5.20}$$

$$X(f) = |X(f)|e^{i\phi(f)} \tag{5.21}$$

where the magnitude and phase of $X(f)$ are given as $|X(f)| = \sqrt{X_{real}^2 + X_{imag}^2}$ and $\phi(f) = \tan^{-1}\left(\dfrac{X_{imag}}{X_{real}}\right)$, respectively.

The magnitude and phase of a complex number are displayed in a bode plot, and the imaginary versus real part of the Fourier transform is displayed in a Nyquist plot.

5.5.2 Autopower Spectrum

The power contained in a signal can be conveniently obtained by determining its autopower spectrum. The autopower spectrum of a signal $x(t)$ is denoted as $S_{xx}(f)$, where

$$S_{xx}(f) = X(f)X^*(f) = X_{real}^2 + X_{imag}^2 \qquad (5.22)$$

$X^*(f)$ is the complex conjugate of the Fourier transform of the signal $x(t)$. The asterisk symbol (*) denotes a complex conjugate operation.

For a transient signal, the frequency spectrum is continuous, and the power of such signals is usually denoted by power spectral density, PSD.

$$PSD(f) = \frac{S_{xx}(f)}{\Delta f} \qquad (5.23)$$

Sometimes, to express the linear spectrum magnitude of the above two quantities, the square roots of $S_{xx}(f)$ and $PSD(f)$ are obtained. The auto-spectrum and power spectral density are both real quantities. Phase has no relevance in autospectrum and power spectral density, since both of these are positive real scalar quantities relating to the energy being carried by a signal.

5.5.3 Cross-Power Spectrum

When there are two signals, $x(t)$ and $y(t)$, present in a system, to estimate the relative phase difference between the two signals the cross spectrum, $S_{xy}(f)$ of the two signals is estimated as per Equation (5.24)

$$S_{yx}(f) = Y(f)X^*(f) \qquad (5.24)$$

The cross spectrum is a complex quantity unlike the autospectrum with a magnitude and phase.

5.5.4 Frequency Response Function

The ratio of the response of the system to its excitation, as shown in Figure 5.9, can be estimated as follows, where $m(t)$ and $n(t)$ are the measurement noise present at input and output, respectively.

$$H(t) = \frac{y(t)}{x(t)} = \frac{v(t) + n(t)}{u(t) + m(t)} \qquad (5.25)$$

Equation (5.25) represents the impulse response function of the system; in the frequency domain it is known as the *frequency response function* (FRF), $H(f)$. The system's frequency response is a complex quantity that

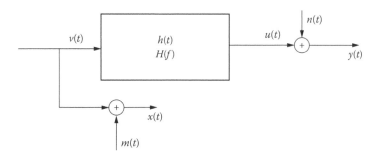

FIGURE 5.9
System with measurement noise in both input and output.

has a magnitude and phase. The phase angle is of much significance as it represents the time delay between the system's response to an excitation.

$$H(f) = \frac{S_{yx}(f)}{S_{xx}(f)} \tag{5.26}$$

When noise is only present at the output, and to minimize its effect in the FRF measurements, the FRF is denoted as $H_1(f)$, which can be estimated by Equation (5.26). When noise is present at the input, and to minimize its effect in the FRF measurements, the FRF is denoted as $H_2(f)$, which can be estimated as

$$H_2(f) = \frac{S_{yy}(f)}{S_{xy}(f)} \tag{5.27}$$

The accuracy of frequency estimates of all signals improves with the number of linear averages. The error of estimates is inversely proportional to the square root of the number of averages.

5.5.5 Coherence Function

In machinery condition monitoring, there are instances when, through measurements and signal processing, one would like to determine the root cause of a problem or eliminate the possibility of the source of the problem. This can very well be identified by determining the correlation between two signals. This correlation can be estimated in the frequency domain and is known as the coherence function, $\gamma_{xy}^2(f)$.

The coherence function estimate is given by Equation (5.28).

$$\gamma_{xy}^2(f) = \frac{\left| S_{xy}(f) \right|^2}{S_{xx}(f) S_{yy}(f)} \tag{5.28}$$

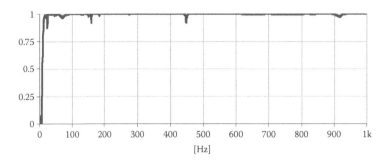

FIGURE 5.10
Measured coherence function between response and excitation.

The range of the coherence function lies between 0 and 1, $0 \le \gamma_{xy}^2 \le 1.0$. A coherence value of 0.0 indicates that the two signals are not at all correlated and are independent of each other. A coherence value of 1.0 indicates a strong cause-and-effect relationship between the two signals $x(t)$ and $y(t)$. The coherence function is extensively used in experimental modal analysis to determine if the response of a structure is only due to its excitation or not. Figure 5.10 shows the coherence function obtained between the transverse vibration response of a cantilever beam and a random force excitation in the range of 0–1 kHz after 100 averages.

5.6 Computer-Aided Data Acquisition

To perform digital signal analysis on a signal measured by a transducer mounted on machinery, signal data needs to be collected. The signal measured by the transducer is analog in nature and continuous in time. The data is collected by a device known as the *analog-to-digital convertor* (ADC) and the digital samples x_i are stored in a memory space in the device for subsequent digital computation. We will not discuss the hardware aspects of the ADC convertor, but we will focus attention on the features of the ADC conversion system of which a machinery fault diagnostics person should be aware. The purpose of the data acquisition system is to accurately represent the measured analog signal to its corresponding digital values. The following two important aspects of the data acquisition system need to be considered, among others:

- Sampling frequency
- Digital bit size

The data has to be sampled at an adequate sampling frequency. In Figure 5.11, the original sample adequately sampled by a sampling frequency

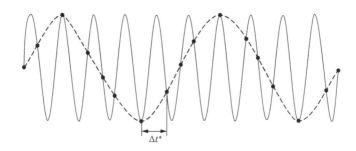

FIGURE 5.11
Signal aliasing.

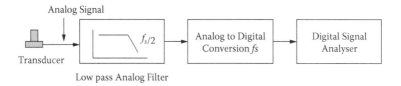

FIGURE 5.12
Configuration of data acquisition system for high-frequency signals.

of $f_s = \dfrac{1}{\Delta t}$ is represented by the dark line. In the same figure, the dashed line indicates the same signal sampled at a slower sampling rate, $f_s^* = \dfrac{1}{\Delta t^*}$. Due to the slower sampling rate, the original signal appears to be a low-frequency signal. In other words, the signal has been aliased as a low-frequency signal. This is a serious error in the data acquisition system, and is known as the *aliasing error*. To prevent signal aliasing, the signal has to be sampled at a rate at least two times higher than the maximum frequency of the signal present in the system. This fact is stated in *Shannon's sampling theory*. When data acquisition is done for a signal whose maximum frequency is not known, a low-pass analog filter with a cutoff frequency of $\dfrac{f_s}{2}$ is used to prevent signal aliasing, as shown in Figure 5.12. This is a very important fact to consider when using data acquisition for dynamic signals like noise and vibration that change very quickly. Data acquisition devices without the low-pass anti-aliasing filters are available for acquiring static signals over a period of time, for example, the temperature signals from thermocouples. When there are multiple inputs to the ADC, each known as an input channel, the sampling frequency is expressed as the number of data points per second per channel. A digital switch known as a *multiplexer* is used to routinely scan the channels in a sequence for data acquisition by an ADC.

Another important point is the corresponding digital value assigned to a digitized analog signal by the ADC. The ADCs store the digital data in binary bits as a digital value corresponding to 2^n, where n is the bit size of the ADC. ADCs are available in many bit sizes of $n = 3, 10, 12, 24$, etc. The maximum analog voltage to a digital convertor is ±5 V or a range of 10V. Thus, the minimum analog that can be detected by an ADC is given by Equation (5.29) as

$$\text{Amplitude Resolution} = \frac{Range}{2^n} \qquad (5.29)$$

For example, for a 3-bit ADC, there are a maximum of 8 digital values from 000 to 111, that can be used to digitally represent the input analog voltage of 10V range. This corresponds to an amplitude resolution of 1.25 V as per Equation (5.29). The problem arises when the analog voltage at a particular instance is at an amplitude resolution less than 1.25 V. Thus the small voltage variations in the signal less than the amplitude resolution of the ADC cannot be captured. This is shown in Figure 5.13 and is known as the *digitization error*. To overcome the digitization error in the ADC process, it is advisable to have an ADC of a higher bit size. For example, for the same input voltage range of 10 V, with a bit size of 12, the amplitude resolution would correspond to 2.49 mV. Thus, very small deviations in the analog signal can be captured by the ADC conversion process. Many times during data acquisition, analog amplifiers are used before the ADC process to amplify the signal so that the ADC device can capture the small changes in the analog signal. The input analog signal to the data acquisition system can be unipolar, where the signal is referenced to a ground voltage using a single wire system, or it can be bipolar with reference to a high and low value of the analog signal. The noise associated with the data acquisition process reduces with the bipolar input.

The digital data thus obtained by the ADC process need to be stored in the digital memory for further computations. Depending upon the number

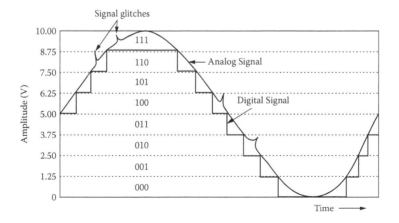

FIGURE 5.13
Effect of bit size on digitization.

of lines of FFT required, the data size, N, will change. Some ADC devices have on-board random access memory (RAM) to store the digitized data. The data acquisition process is controlled by driver software that is resident on the host computer wherein the triggering of the data acquisition process can be initiated, based on a certain input voltage level. The software can also control the rate at which the data is stored in the on-board memory, the mode and time of data acquisition, and whether it should be in a continuous mode or intermittent. Many standard commercial hardware systems are available for data acquisition along with their driver software. The digital data thus acquired, needs to be transferred to the computer system through a data transfer protocol, based on the architecture of the computer. Thus the compatibility of the ADC hardware with the computer system must be ensured. Some of the standard computer architectures over the years that have been used for interfacing with the ADC are ISA (Industry Standard Architecture), EISA (Extended Industry Standard Architecture), PCI (Peripheral Component Interconnect), PCMCIA (Personal Computer Memory Card International Association), and USB (Universal Serial Bus).

Nowadays, data acquisition devices are available that can transmit digital data over a wireless network, and the resident computer can be in a local area network (LAN) or via Ethernet with other computers on the Internet and share data with signal analysts at locations far away from the actual machine where the analog data is captured. Even aircrafts in flight transmit data to the ground maintenance group located at a remote location on the globe who monitor the real-time aircraft engine health conditions by monitoring critical parameters using data transmitted to them through satellite-based communication systems.

5.7 Signal Conditioning

The signal from the transducer on a machine may require additional processing, like signal amplification, noise reduction, filtering, linearization, and so on. These functions are usually done through standalone analog signal conditioners, and sometimes some of these functions are done in the digital domain through dedicated digital signal processing software after the analog-to-digital conversion. Some of the transducers require an external power supply, which could be provided by the signal conditioners. A common requirement is to supply 4-mA current to many of the integrated charge–type noise and vibration transducers.

5.7.1 Signal Filtering

During signal processing, a requirement arises to analyze the acquired or measured signals in a particular frequency band of interest. This is achieved by filtering the signals. Signal filtering can be done both in the analog domain and the digital domain. Following are the common analog filters used in signal processing:

i. high-pass filter
ii. low-pass filter
iii. band-bass filter
iv. notch filter

A high-pass filter allows signals with frequencies beyond a cut-on frequency to be passed through. Usually in machinery condition monitoring, high-pass filters are used to remove near-mean or DC values of the signal, and cut-on frequencies of 0.1 Hz or 1 Hz are quite common. Another very important filter is the low-pass filter, and its significance in preventing signal aliasing during data acquisition was mentioned earlier. Low-pass filters allow only signals to pass up to a cutoff frequency. A combination of a high-pass and low-pass filter can be used as a band-pass filter, which allows only signals in a particular frequency band to be passed through. Many times, due to a ground loop with the electrical supply frequency, the electrical supply frequency (50 Hz or 60 Hz) shows up in the acquired machinery signals. This single frequency can be removed by using a notch filter. It may be mentioned here that the electrical supply frequency in some European and Asian countries is 50 Hz, whereas in the Americas it is 60 Hz. At the cutoff and cut-on frequencies, the filters are not sharp and some roll off occurs, which depends on the order of the filter.

Figure 5.14 shows the frequency response of a high-pass filter with a cut-on frequency of 200 Hz, a low-pass filter with a cutoff frequency of 200 Hz, and a band-pass filter of frequency band 200 Hz to 1 kHz. Figure 5.15 shows the time waveform of a random noise signal of 0–2 kHz, and the respective filtered time histories of the signal after passing through the high-pass filter only, the low-pass filter only, and the band-pass filter only.

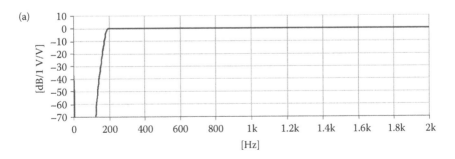

FIGURE 5.14
Frequency response of filters: (a) high-pass filter.

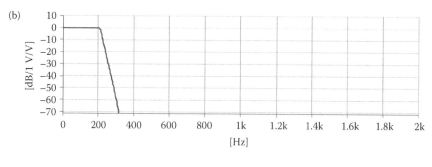

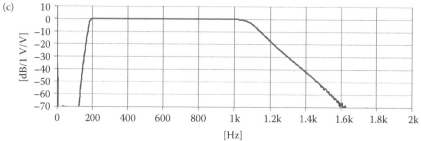

FIGURE 5.14 (*Continued*)
Frequency response of filters: (b) low-pass filter, and (c) band-pass filter.

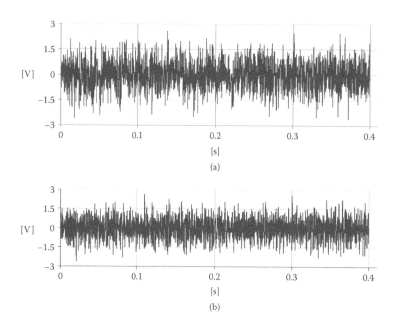

FIGURE 5.15
Time histories of the random signals: (a) 0–2 KHz random noise signal; (b) 200 Hz high pass filtered.

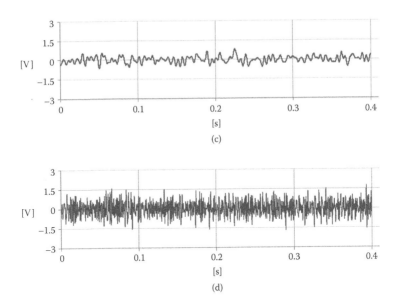

FIGURE 5.15 (*Continued*)
Time histories of the random signals: (c) 1 KHz low pass filtered; (d) 200 Hz–1 KHz band pass filtered.

5.8 Signal Demodulation

When the amplitude or frequency of a signal varies with time, it is called a *modulated signal*. For modulated signals, spectral analysis using direct FFT is not adequate to extract the required features. And in such cases, to extract the modulating signal, demodulation is necessary. Equation (5.30) (a and b) give the expression for an amplitude modulated wave. The sidebands around the modulating frequency indicate the modulations.

Demodulation is mostly used in communication systems where shifting of a frequency bandwidth by a high-frequency carrier plays an important role in signal transmission. When the signal is transmitted, it is mixed with another high frequency sinusoidal signal (i.e., carrier signal), which is termed modulation and reverse (i.e., the separation of the carrier signal) and is known as *demodulation*.

$$x_{am}(t) = A\left[1 + k_{am}\cos(2\pi f_m t)\right]\cos\left(2\pi f_c t\right)$$

$$x_{am}(t) = A\cos\left(2\pi f_c t\right) + \frac{Ak_{am}}{2}\left\{\cos\left[2\pi(f_c + f_m)t\right] + \cos\left[2\pi(f_c - f_m)t\right]\right\}$$

(5.30) (a-b)

where K_{am} is the index of amplitude modulation.

Figure 5.16 shows an amplitude-modulated (AM) wave with a carrier frequency of 20 Hz and modulated frequency of 5 Hz. The FFT of the AM wave in Figure 5.17 shows the sidebands around the carrier frequency.

By using the Hilbert transform, the envelope of the AM signal can be obtained as shown in Figure 5.18. The FFT of the envelope in Figure 5.19 clearly shows the modulated signal of 5 Hz.

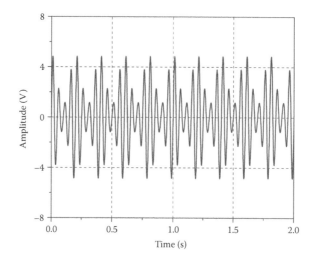

FIGURE 5.16
Amplitude modulated signal.

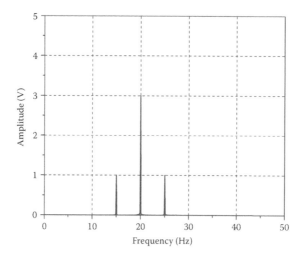

FIGURE 5.17
Spectrum of amplitude-modulated signal.

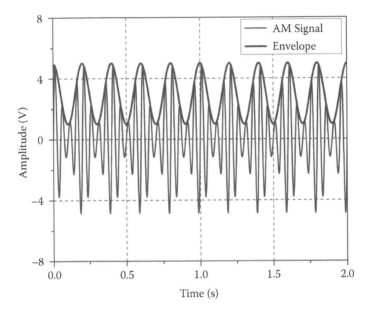

FIGURE 5.18
Envelope of modulated signal.

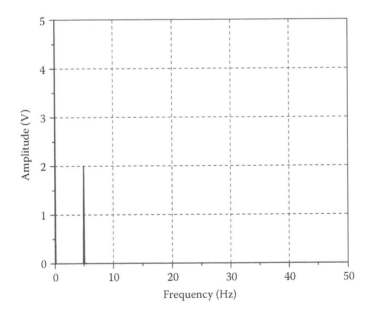

FIGURE 5.19
Spectrum of the modulated signal.

5.9 Cepstrum Analysis

When the families of sidebands in signals become many and the signals have a poor signal-to-noise ratio, cepstrum analysis is used to recover such signals in the time domain. This is particularly used in machinery condition monitoring for gearbox fault diagnosis. Equation (5.31) gives the expression for the cepstrum of a signal.

$$Cp(\tau) = F^{-1}\{\log F_{xx}(f)\} \tag{5.31}$$

where F^{-1} is the inverse Fourier transform of the signal.

5.10 Examples

5.10.1 Natural Frequency of a Cantilever Beam

To illustrate the representation of signals in the frequency domain, examples from the experimental modal analysis of a cantilever beam are presented here.

Natural frequencies of a cantilever beam (one end fixed and other end free) are studied. The beam is rectangular in cross section and the dimensions are length (L) 30 cm, width (b) 20 mm, and thickness (t) 3 mm. Material properties are as follows: Young's modulus (E) 200 GPa and density (ρ) 7800 kg/m^2. Natural frequency can be computed by Equation (5.32)

$$f_{theory} = \frac{1}{2\pi}\sqrt{\frac{EI}{\rho A}}\left(\frac{p}{L}\right)^2 \tag{5.32}$$

where I is the area moment of inertia and $I = \dfrac{bt^3}{12}$. p is a frequency parameter shown in Table 5.4.

Figure 5.20 shows the H1 FRF magnitude obtained from the experimental modal analysis. Figure 5.21 (a and b) show the real and imaginary parts of the H1 FRF. Figure 5.22 shows the Nyquist plot of the FRF.

TABLE 5.4

Natural Frequencies of a Cantilever Beam in Hz

p	1.875	4.694	7.855	10.996
f_{theory}	27.3	170.9	478.5	937.6
$f_{experiment}$	23.44	159.7	449.1	914.1

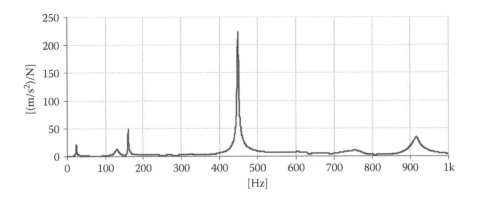

FIGURE 5.20
H1 FRF magnitude spectrum.

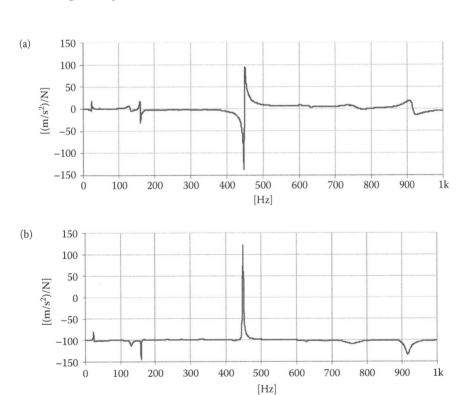

FIGURE 5.21
(a) Real and (b) imaginary spectrum of the H1 FRF.

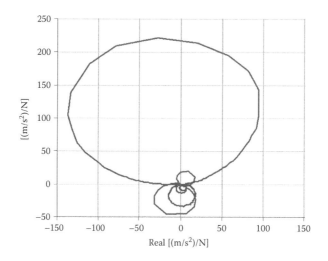

FIGURE 5.22
Nyquist plot of the FRF.

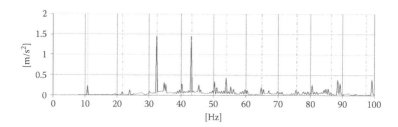

FIGURE 5.23
Acceleration spectrum of compressor vibration.

5.10.2 Compressor Vibration

Figure 5.23 shows the acceleration spectrum of the vibration signal of a reciprocating compressor operating at 650.5 RPM (10.84 Hz). The harmonics of the rotational frequency are evident in the acceleration spectrum. The FFT corresponds to 3200 lines with a frequency resolution of 0.156 Hz. The overall vibration level of the compressor in the frequency range of 0 to 500 Hz is 3.609 m/s². The velocity spectrum is shown in Figure 5.24.

5.10.3 Engine Vibration

An optical encoder (described in Chapter 6) was used to measure the rotational speeds of an engine and detect its firing frequency. It is assumed that the engine speed is a constant, but there are variations around 2100 RPM, as shown in Figure 5.25. For signals where the rotational speed varies, it is

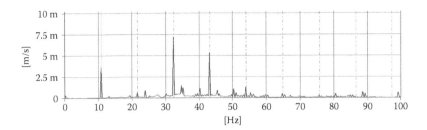

FIGURE 5.24
Velocity spectrum of the compressor vibration.

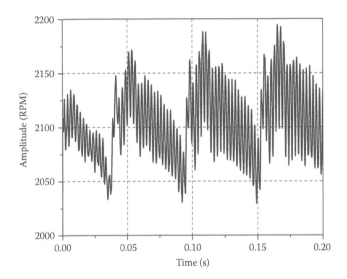

FIGURE 5.25
Instantaneous rotational speed of an engine.

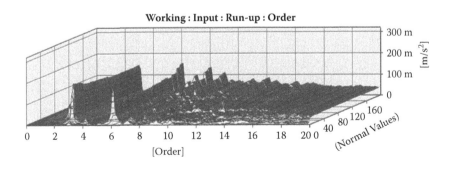

FIGURE 5.26
Unbalance order in a rotor rig.

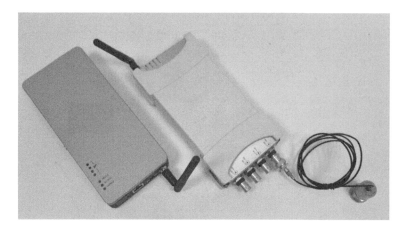

FIGURE 5.27
An wireless data-acquisition device.

best to use a trigger that is synchronized with the shaft rotation. An order is defined as the rotational speed of the machine. Figure 5.26 shows the vibration of a rotor having an inbalance as an order plot. In places like an engine in a remote location that is not accessible for acquiring data, acquired data can be transmitted over wireless. Figure 5.27 shows a wireless data acquisition with its transmitter and receiver.

6

Instrumentation

6.1 Introduction

The signals from machines need to be measured and analyzed to interpret the condition of the machine. To achieve this, the machine should first be instrumented with the appropriate transducer to measure the mechanical parameter that manifests as a signal from the machine. In machinery condition monitoring, some of the most common mechanical parameters measured are vibration, temperature, motor current, oil condition, and so on. In this chapter, a brief insight into the principles of measurement and the appropriate transducers required for measurement of the common mechanical parameters are provided. A brief on the working principle behind the transducers, their special features, and applications is also provided.

6.2 Measurement Standards

Worldwide standards are established for the measurement procedure to be followed for measuring a particular parameter. These standards also have their regional and national versions. A list of such standards is provided in Appendix A6. When a standard is adopted, it becomes easier to communicate the data obtained at one machine location to another location in another part of the world where the data may be analyzed by another different organization altogether. Such standards also help the equipment manufacturer, transducer developer, and maintenance engineer, so that they all have the same common platform to share their concerns.

6.3 Measurement Errors

Despite the best efforts and intentions of the maintenance engineer in the field, errors in measurement creep in. This could be because of several reasons, such as lack of knowledge about the measuring equipment, incorrect installation of the transducer and its accessories, using the transducer to measure a quantity it is not designed to measure, and so on. Efforts are always made to measure the physical quantity most accurately, because the interpretation of the machine quality depends on the measured data. Measurement errors can be broadly classified into two categories, random error and bias error. Random error in a measurement can be reduced by performing an arithmetic average of the measured results. This means that a number of measurements need to be done. Bias error or offset error is due to a fixed amount of difference between the actual quantity and the measured quantity. This shift or difference is usually known as the *offset* or the *bias*. Bias error can be reduced by comparing the measured quantity with the actual quantity, which should be present. The actual quantity is determined by calibration against a known reference quantity. There are instruments that are precise, and such instruments, in repeated measurements of the same physical quantity, always produce the same output. An ideal measurement must thus be both precise and accurate, with no errors.

The smallest mechanical quantity that can be measured is the least count that is available on the display of the measuring equipment. The uncertainty in a measurement is one half of the least count available on the measuring equipment. Thus, when selecting an instrument for measurement, one must be aware of its accuracy, precision, and least count.

6.4 Calibration Principles

Instruments can be calibrated in several ways. The most common method of calibration is to compare with a reference standard that is considered to be more accurate and precise than the instrument being calibrated. Usually, in practice or in test laboratories, it is a good practice to have one set of the same transducers (known as a laboratory reference) kept apart from other transducers used in day-to-day field measurements. It is a good practice to compare the readings of the laboratory reference with the remaining transducers at regular intervals or when there is a need to cross check both readings. The laboratory standard or *tool room standard* is in turn calibrated against a better regional or national standard maintained at the accredited test laboratories.

Many times in the field while performing a measurement on a machine, there are instances in the measurement chain like amplifiers whose gain settings are not known, or even instances where the sensitivity of the installed transducer is not known. Usually in such cases a portable mechanical reference source is used, whose output mechanical parameter is known. For instance, an ice-water bath in a beaker at atmospheric pressure is known to be at 0°C. Such a bath can be used to calibrate thermocouples that are used for temperature measurement. Commercially, such calibrated reference sources are available for vibration, sound, strain, oil chemical composition, and so on.

6.5 Static and Dynamic Measurements

There are two situations in the output of a mechanical system—one in which the mechanical quantity being measured changes quite quickly with time, and another in which the quantity hardly changes with time. The temperature in a room is of the former type and the vibration of an engine is of the latter type. Static measurements are done for cases when the mechanical quantity does not change with time, whereas dynamic measurements are for the case when the mechanical quantity changes with time. In static measurements, it suffices to measure once in a while and to interpret the measured values to be constant with time. For dynamic measurements, one needs to measure within the time it takes for the mechanical quantity to change in order to draw any meaningful conclusions. In dynamic measurement, the inertia, stiffness, and damping of the transducer play an important role. Care must be taken that the above characteristics of the transducer do not influence the dynamic output of the mechanical system.

6.6 Frequency Response

In dynamic measurements, frequency response of the transducer plays an important role. Every transducer can be considered as a mechanical system with inertia, stiffness, and damping characteristics. Such transducers will thus have a natural frequency of their own, for a single-degree-of-freedom system representation of the transducer, its damped natural circular frequency in rad/s is given by Equation (6.1)

$$\omega_d = \omega_n \sqrt{1 - \xi^2} \tag{6.1}$$

where $\omega_n = \sqrt{\dfrac{k}{m}}$ and ξ is the damping factor.

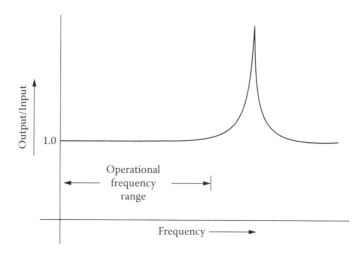

FIGURE 6.1
Frequency response curve of a transducer.

While performing dynamic measurements, it is desired that the frequency of the measured quantity is nowhere close to the natural frequency of the transducer. Thus, every transducer is provided with an operational frequency range, and this frequency range is different from the natural frequency of the transducer system. A typical frequency response curve of a transducer is shown in Figure 6.1.

The natural frequency of the transducer is high for less inertia and more mounting stiffness. These are to be considered while selecting a type of transducer for dynamic measurements. The abscissa of the plot in Figure 6.1 shows the ratio of the transducer output to the mechanical input. For a good measurement frequency range, this ratio should be unity or 0 dB (since the logarithm of 1 is 0). At the natural frequency of the transducer, the output is amplified and leads to a wrong conclusion of the mechanical parameter being measured. Usually, measurement at a frequency within a region of 10% around the natural frequency of the transducer is to be avoided. Transducer manufacturers usually provide the individual measured frequency response along with the instrument.

6.7 Dynamic Range

Another important characteristic of a transducer is its dynamic range, which relates to the maximum and minimum quantity that can be measured. The dynamic range is usually expressed in a dB scale and is the natural logarithm of the ratio of the maximum to minimum quantity measured by the transducer. A typical dynamic range of a transducer can be anywhere from 60 to 100 dB. This dynamic range is also a function of the frequency of the measured signal.

6.8 Basic Measuring Equipment

In condition monitoring of machines, the mechanical parameters that are measured as a signal by the transducers many times need to be filtered, amplified, and displayed on an analog or a digital meter. Sometimes the measured signal needs to be displayed as a time signal, so that the signal features can be noticed, which helps to make an in situ decision in the field if the acquired signal is of use or not. A few basic instruments are used in the field or in the laboratory for a preliminary study of the signals and some of them are described here.

6.8.1 RMS/Peak Meters

Conventional analog voltmeters and ammeters based on the moving iron principle are analog in nature and widely used. These analog instruments are used for both AC and DC signals. Suitable rectification and analog operations are done on the AC signal to obtain its root mean square (RMS) value. The RMS value of a signal is a measure of its heating ability for a pure resistive load. For the case of a harmonic signal, the peak value of a signal is $\sqrt{2}$ times its RMS value. Nowadays, meters are available with digital readouts or displays where an analog-to-digital convertor (ADC) is used to convert the signal, and then the corresponding digital values are displayed in a digital LED readout.

6.8.2 Signal Amplifiers

The signals from the transducers at times need to be amplified so that they can be detected by the ADC. Many times a signal needs to be amplified so that it can drive an actuator. Amplifiers are used to perform these activities. An operational amplifier (Op-Amp) is used to do some of these activities. An Op-Amp is used to provide a high gain to the input signal. An Op-Amp has other important functions in signal conditioning, like inverting the input, integrating or differentiating an analog, adding a few signals, and so on. Op-Amps usually require a stable regulated power supply to operate.

6.8.3 Oscilloscope

This is an instrument where the acquired voltage signal from a transducer can be displayed as a function of time. Earlier oscilloscopes were analog in nature and worked on the principle of a cathode ray tube (CRT). However, with the digital revolution, digital oscilloscopes are popular. In a digital oscilloscope, an analog-to-digital converter (ADC) is used to convert the measured voltage into a digital signal. Some of the popular digital oscilloscopes are the digital storage, digital phosphor, mixed signal, and digital sampling oscilloscopes. Digital oscilloscopes usually have the ability to acquire the signal at a sampling frequency selected by the user, and display on a screen the digital data as a function of time. Some oscilloscopes can acquire more than one signal at a time and perform simple arithmetic operations on the acquired digital signals. Provision of AC and DC coupling of the waveform, and display of the RMS and mean value of the signal are also available. Nowadays, the digital oscilloscopes available also have additional memory for storage of the data or transfer to a computer through a USB. The acquisition of the signal in a digital oscilloscope can be acquired by a user-set trigger level, based on the amplitude of the signal. Provisions to observe the pretriggered data are also available. Digital oscilloscopes with high sampling rates are convenient to quickly capture changing transient events, for example, the vibration due to a ball passing over an inner race defect in a ball bearing. The least voltage displayed in a digital oscilloscope is dependent on the resolution of the ADC. Today, 24-bit ADCs are quite common in digital oscilloscopes. Since a CRT is not present in digital oscilloscopes, they are quite light and portable.

6.8.4 Signal Generators

In several instances, to excite a device under test (DUT), an input energy in the form of a signal is given. Depending on the amount of the input energy, the signal may be amplified by a power amplifier to drive the actuator. Different tests require different types of input excitation signal. For instance, to determine the natural frequency of a system, a sine sweep signal or a random noise signal of a certain frequency bandwidth is desirable. Many times, to determine the system characteristics of a DUT, a sine wave, step, or ramp input is preferred. Signal generators are devices that provide a user-defined signal of a particular voltage level and frequency or time period. Earlier, signal generators were essentially analog oscillators tuned to a particular frequency. However, again, with the advent of digital electronics, the present-day signal generators are digital and they produce an analog signal using an onboard digital-to-analog converter (DAC). Some of the most common signals produced are sine, square, sawtooth, haversine, random, and pink noise. In many signal generators, there is also a provision to produce a user-defined or arbitrary time waveform.

These are particularly helpful in durability tests on components by exciting them with measured real-world data. Some of these types of signal generator are known as *arbitrary waveform generators*. There is also capability to produce amplitude- and frequency-modulated signals on a particular carrier frequency. Another feature available in such signal generators is to weigh the output signal with a DC bias to stimulate a static preload condition in the DUT. Such generators can be used to calibrate dynamic transducers as well.

6.8.5 Signal Filters

Both analog and digital signals need to be filtered. Analog filtering is done by a suitable resistor–capacitor (RC) circuit, and such filters can be designed to be of a high pass or a low pass type. Passing a signal through a high pass at first, which is then followed by a low-pass filter, enables one to have band-pass filtering on the signal. Anti-aliasing low-pass analog filters are used before any analog-to-digital conversion, to prevent aliasing of a signal whose maximum frequency content of the signal is not known. These filters usually have the provision to AC- or DC-couple the signal with preamplification when necessary. Analog filters have a roll-off since the filters are not sharp, and a roll-off of −60 dB/decade is common. In order to have sharp roll-off, use of digital filters are convenient. Such digital filters can be applied to digital signals obtained after ADC processing. These digital filters can be realized through a computer program or a programmable digital signal processor (DSP). The advantage of a DSP is that any filter can be realized, and they can be adapted with proper selection of the filter coefficients. Very high filter roll-offs are possible with digital filters.

6.8.6 Power Supply

A power supply is required to power equipment like amplifiers, filters, transducers, and so on. Some of the portable power sources are dry cell batteries like alkaline cells, lead acid batteries, nickel-metal hydride (NiMH) batteries, and lithium-ion (Li-ion) batteries. Nowadays, Li-ion batteries are preferred for their high power-to-weight ratio and long life. These batteries are portable and can be used for field use to power the condition monitoring equipment. However, in the laboratory or test facility, an electrical AC supply based power supply source is preferred. These power supplies essentially have a step-down transformer with a notch filter to remove the electrical supply frequency, followed by rectification to provide a DC output voltage. Depending upon the secondary voltage tapping of the step-down transformer, the appropriate DC voltage can be selected.

6.8.7 Counters

In condition monitoring, there are many mechanical events where the number of such events is of relevance. Counters are used to measure the number of occurrences of such repetitive events. The counters usually have a clock set at a particular frequency, where in a given time period that is the reciprocal of the clock frequency, events are counted using digital registers so that the number of such events in a given time period is a measure of the frequency of the event. Such frequency counters are very useful for determining the rotational speed of the machine being measured by an inductive probe, positioned near the rotating shaft.

6.9 Vibration

Vibration is one of the most common mechanical parameters of a machine that is measured for condition monitoring purposes. Vibration can be measured in terms of its displacement, velocity, or acceleration. In this section, the popular transducers that are used to measure vibration are described.

6.9.1 Displacement

For slowly rotating shafts or shafts at rest, many times displacement is measured. During static conditions, relative displacement of the shaft with its housing is measured by a simple linear distance measurement device like a dial gauge. At slow speeds, the displacement is measured by a linearly varying displacement transducer (LVDT). These LVDTs have a frequency response of at most about 50 Hz.

A proximity probe based on the eddy current principle can be used for measuring displacement, however this probe requires an additional high-frequency AC signal, and the surface whose displacement is being measured must be electrically conductive. Another commonly used displacement measuring device is a capacitive probe, which has a wide frequency response, but here again the surface must be electrically conductive. In some instances, position potentiometers are used for displacement measurements, whose dynamic range and frequency response are limited.

6.9.2 Velocity

Linear vibration velocity is measured by a self-generating low-impedance vibration velocity transducer. However, such transducers are severely limited in their frequency response and dynamic range. Because of the moving coil, these transducers have a low natural frequency. Hence, they are usually not suitable for measurements of vibration velocity less than 10 Hz.

6.9.3 Acceleration

Acceleration is measured based on the principle of the relative motion of a suspended mass in a casing, where the base of the casing is subjected to a motion. By measuring the motion of the suspended mass, the acceleration of the base can be measured. A sensing element is attached to this suspended mass, and the output is calibrated to provide a measure of the base acceleration. Today, piezoelectric crystals are used as a sensing element, though in the past, strain gauge–based accelerometers were used.

In a piezoelectric accelerometer, the piezoelectric element (which is the sensing element) is placed between the base of the accelerometer and the top of a mass on the accelerometer. The advantage of using piezoelectric crystals is that they do not require any external power supply and are relatively stable at temperatures as high as 400°C. When compressed, the piezoelectric crystals produce an electrical charge, and with appropriate conversion of this charge to a voltage, the acceleration can be measured and recorded. Another important characteristic of the piezoelectric crystals is that they are sensitive to motion in a particular direction. Since vibration is measured along a direction, the piezoelectric crystal is aligned along the most sensitive axis inside the accelerometer housing. Along with a piezoelectric accelerometer, a preamplifier is required, to convert the high-impedance output into a low-impedance voltage signal for direct transmission to measuring or analysis equipment like an oscilloscope or a dynamic signal analyzer. This preamplifier is known as a *charge amplifier*.

A charge amplifier uses an operational amplifier at the input stage. The configuration of the operational amplifier with the capacitor in the feedback loop operates as an integration network and integrates the current at the input. This input current is the result of the charge developed across the high-impedance piezoelectric elements inside the accelerometer. The amplifier works to nullify this current and in doing so produces an output voltage proportional to the charge. The voltage output of a charge amplifier is proportional to the charge at the input, and therefore to the acceleration measured by the accelerometer. The low-frequency response of a charge amplifier is determined by the time constant set by the feedback circuit around the operational amplifier and is unaffected by changes in the input load conditions. The lower limiting frequency is changed by varying the feedback resistance. The sensitivity of a charge amplifier is not significantly affected by the change in capacitance caused by changing cable lengths. When very long cables are used, the high-frequency response is slightly attenuated. The use of very long accelerometer cables and low-gain settings will increase the noise of the charge amplifier and hence reduce the signal-to-noise ratio of the measurement. If the resistive load at the input drops significantly, the noise will also increase. There are two predominant sources of noise in a charge-type accelerometer—one is the triboelectric noise cable, and the other is ground-loop noise. The triboelectric noise cable will generate a charge if its mechanical movement is not restricted. Thus, the cables used in such accelerometers are layered and reinforced, so that no sliding of the cables occurs. Ground-loop noise occurs in machines where there is a possibility that the machinery housing may not be at earth potential. Consequently, the accelerometer case and cable screen will not be at earth potential and a voltage drop will exist along the cable. In some instances, electromagnetic interference (EMI) might also contribute to the measurement noise. Thus, in practice, coaxial cable is preferred since such EMI noise can be minimized. During measurement, care is to be taken to ensure that the accelerometer cables are not running alongside the power cables.

There are instances when the Op-Amp circuit used for the charge-to-voltage conversion is kept inside a hermetically sealed accelerometer casing and is provided with a power supply of 4 mA constant current from an external power source. Such accelerometers are convenient since the special triboelectric reducing cables need not be used, and instead of the charge amplifier, only the power supply is required. Such accelerometers are quite popular and are available on the market under various trade names like ICP type, Piezotron, Isotron, and so on. However, the greatest disadvantage of accelerometers with built-in charge amplifiers is that because of the electronics, they cannot be subjected to high temperatures of more than 200°C for long periods of time; the electronic components inside the accelerometer casing will get damaged beyond repair.

The piezoelectric accelerometer used for vibration measurements is fixed to a surface whose vibration needs to be measured. Some of the common accelerometer mounting techniques are:

  i. beeswax
 ii. adhesive
iii. magnet
 iv. stud
  v. handheld probe

For permanent monitoring of an accelerometer, it is usually stud mounted or glued to the surface. The surface must be cleaned and polished so that there is a uniform and firm contact with the accelerometer base. For quick attachment of miniature accelerometers at room temperature, beeswax is preferred. Many prefer magnets for attaching an accelerometer to the machine surface, since they quickly provide a very rigid mounting. However, for magnet mounting, the surface should attract a magnet. Care must be taken while mounting an accelerometer using magnets at high temperature, since a magnet made of a soft material may develop a crack and fail at high temperatures. To prevent an electrical ground loop, a mica washer is introduced between the machine surface and the accelerometer base. For a quick vibration survey, a handheld probe is attached to the accelerometer; however, the frequency response of this mounting method is restricted to only about 1 kHz.

Accelerometers that are used to measure vibration in only one direction are known as *uniaxial accelerometers*. However, during machinery vibration monitoring, the vibrations in all the three linear directions need to be measured. In such situations, rather than using three uniaxial accelerometers, a single accelerometer with three piezoelectric elements in three mutually independent directions is placed inside the accelerometer casing, with three terminals for signal cable connections. These accelerometers are known as *triaxial accelerometers*, and they can also be mounted by the techniques described earlier. A variety of accelerometers used for condition monitoring are shown in Figure 6.2(a). A handheld calibrator that provides an acceleration level of 10 m/s² at 1000 rad/s is shown in Figure 6.2(b).

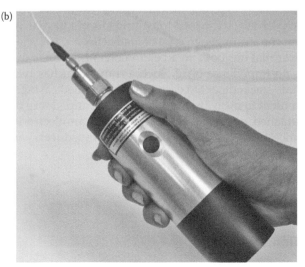

FIGURE 6.2
(a) Accelerometers used for vibration measurement, (b) A handheld calibrator for accelerometer.

The charge sensitivity of the accelerometers are usually given in pC/ms^{-2} and after a suitable voltage conversion by the charge-to-voltage amplifier, the output is provided as mV/ms^{-2}. These amplifiers have a few other features, like signal cable fault detection, signal overload detection, filters to set the upper and lower frequencies of measurement, amplification to a certain gain, and an integrator for converting the measured acceleration to either voltage or displacement. In the field, due to use of long cables, there may be a loss of signal strength and it is usually a good practice to use an in situ handheld calibrator to calibrate accelerometers.

6.10 Force Measurements

In many applications, particularly during component testing, the measurement of dynamic force is required. In such applications, the piezoelectric crystals are once again used. These materials are usually given a precompression and held between discs in a sealed unit. When this transducer is subjected to both tensile and compressive forces, it produces a proportional AC signal. These force transducers also require a charge amplifier for signal condition. In tool condition monitoring for measuring cutting forces, piezo-based cutting force dynamometers are extensively used. The typical range of such force transducers are from −12 kN to +10 kN.

For very large force measurements, load cells with strain gauges are used. However, such strain gauge load cells require an additional bridge circuit (like a Wheatstone bridge) for measuring the strains induced by mechanical load. The piezoelectric-based force transducers have a better frequency response than the strain gauge type of load cells.

A transducer that has both an accelerometer and a force measuring gauge is known as an impedance head. Such a transducer has two sets of piezo-electric crystals stacked next to each other in a sealed casing. An impedance head is convenient to measure the driving point impedance of the DUT. This is usually attached on top of the stinger from an electrodynamic shaker. The impedance head can have charge-type transducers, and would thus need a dual-channel charge-type amplifier.

6.11 Rotational Speed

Rotational speed is a very important parameter to be measured in the case of rotating machines, since the dynamic motions of machines are related to their rotational speed. There are many transducers available for measurement of rotational speed. Usually, all of them are of the noncontacting type barring the mechanical tachometer or the tachogenerator. In many rotational speed measurements, a signal that occurs once per rotation of the shaft is timed to determine the rotational speed. However, when the speed of the rotating shaft changes or fluctuates within even rotation of the shaft, an optical encoder with a resolution of 1000 pulses/revolution or better is preferred.

6.11.1 Stroboscope

A stroboscope essentially consists of a high-intensity xenon or krypton lamp that is rapidly switched on and off by an oscillator signal, where the frequency of flashing of the light on the object can be varied. Particularly when the frequency of the light flashing frequency coincides with the speed of the object, the object appears to be stationary. Thus the speed of the object can be known. There are instances when the objects also appear to be stationary even at multiples of their running speed. In such cases, the accurate speed would be the lowest frequency of the strobe light when the object appears to be stationary. Stroboscopes have another application in machinery fault detection. For instance, if a belt is slipping on a pulley, when the strobe light is flashed on the rotating pulley and the belt, the pulley appears to be stationary and the belt appears to slide or slip over the pulley.

6.11.2 Inductive Probe

An inductive probe essentially consists of a coil on a soft iron core kept inside a casing. If such a probe is brought near a moving iron surface, a voltage is induced in the coil. The voltage induced is dependent on the relative position between the coil and the iron surface. Thus, on a rotating shaft where a toothed wheel is mounted, there will be as many pulses per rotation of the shaft as there are teeth present in the toothed wheel. By measuring the time period of the pulses, the time taken for one rotation of the shaft can be obtained, and thus the rotational speed can be estimated. There are many variations of this kind of setup in industries where rotational speeds of machines are routinely monitored. At times, even the key on a shaft can be used to provide a once-per-revolution voltage signal. A frequency counter can also be connected to the inductive probe to display the rotational speed of the shaft. For many data acquisition purposes, such probes are also used to trigger the acquisition. This is particularly useful for time domain synchronous averaging of signals from rotating machines. A commercial name for this signal is the Keyphasor signal.

6.11.3 Optical Tachometer

Another popular rotating speed measurement device is the optical tachometer or phototachometer. In this device, a light from an optical diode is focused

FIGURE 6.3
Fiber-optic cable end for an optical tachometer.

onto a rotating shaft. If a reflecting tape is put on the rotating shaft, which should be otherwise dull, the incident light will be reflected, which can be received by a photodiode receiver kept next to the light transmitter. Depending on the number of reflective tapes on the shaft, a corresponding number of pulses will be generated by the phototachometer, which can be measured by a clocked timer and the rotational speed displayed. These phototachometers require a power supply unlike the reluctance pickup. Typically, a 9- to 12-V DC supply is required for generating the light to be incident on the rotating shaft. Care has to be taken that the rest of the rotating shaft is dull in comparison to the reflecting tape, so that any unwanted reflections are avoided. These phototachometers can be mounted on a tripod and positioned near the shaft so that the light is normally incident on the rotating shaft. In hard-to-reach places, fiber-optic cables can be used to transmit and capture the reflected light from the rotating surface for speed measurements as shown in Figure 6.3.

6.11.4 Optical Encoder

Optical encoders are used for measuring rotational speed to very high accuracy. In fact, researchers have recently used rotary optical encoders to measure the instantaneous angular speeds (IASs) of a shaft, which vary even within the time taken for one rotation of the shaft. These encoders essentially consist of a disc with equally spaced fine slots on the periphery. A light is focused on the slotted disc from one side and a photo-receiver on the other side receives the light as pulses as shown in Figure 6.4. The resolution of the encoder is the number of such slots or pulses/revolution, and resolutions of 500 to 1000 pulses/revolution are quite common.

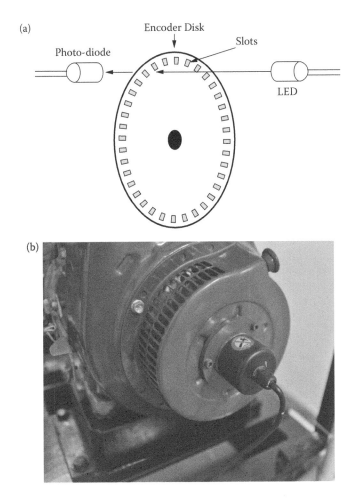

FIGURE 6.4
Optical encoder. (a) schematic representation, (b) used for engine speed measurement.

The basic idea of using IAS signal analysis is that any fault present in a rotating machine will change the rotating dynamics of the machine, and as a result, the speed of the shaft will vary. Therefore, analysis of IAS signals will provide fault-related information on the machine. The encoder provides a pulse signal and the number of pulses in one revolution depends on the resolution of the encoder. A typical pulse signal and estimated IAS are shown in Figures 6.5 and 6.6.

Estimation of the IAS from a pulse signal can be categorized into two groups:

1. Timer/counter techniques
2. ADC-based techniques

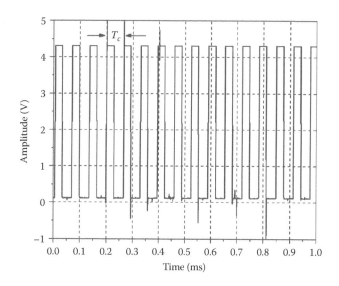

FIGURE 6.5
Pulses from an optical encoder.

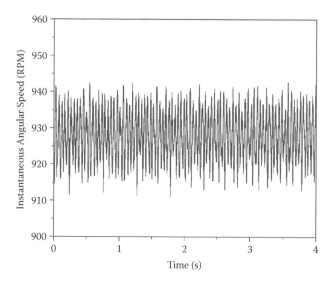

FIGURE 6.6
Estimated instantaneous speed from the encoder pulse.

The timer/counter technique requires two additional devices: a high-frequency clock and a resistor. The high-frequency clock is used as a time reference for measurement of one pulse period. Figure 6.7 shows the pulse signal and corresponding clock pulse signal.

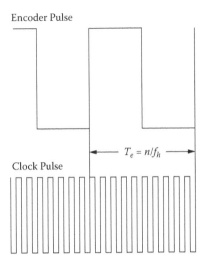

FIGURE 6.7
Encoder pulse signal and corresponding clock pulse signal.

An encoder with M resolutions produces M number of pulses over a revolution. Therefore, the angle covered by each pulse is $2\pi/M$ radian. The high-frequency clock also generates a number of pulses according to its frequency (f_h). As the high-frequency clock is used as a time reference, the time period of an encoder pulse is measured by counting the number of clock pulses corresponding to the pulse duration. If the number of clock pulses per encoder pulse is n, the time period over each pulse is then n/f_h. Further, the elapsed time for each revolution is the product of the number of pulses (M) in one revolution and the time period for single pulse (n/f_h). Therefore, measured instantaneous speed can be written as

$$IAS = \frac{60 f_h}{nM} \text{ RPM} \qquad (6.2)$$

When the speed of a machine increases, the time period of one encoder pulse becomes smaller and the corresponding number of clock pulses decreases. Therefore, the maximum measurable speed should be such that time period of one pulse will be at least the same as one clock pulse time period. Hence, the maximum measurable speed based on the time/counter technique is

$$IAS_{max} = \frac{60 f_h}{M} \text{ RPM} \qquad (6.3)$$

On the other hand, at a lower speed the time period of each pulse becomes higher. Then the number of clock pulses that corresponds to one pulse increases. Hence, the resistor length (bit size) should be such that it must not face resistor overflow. Therefore, the minimum measurable speed based on the time/counter technique is

$$IAS_{min} = \frac{60 f_h}{RM} \text{RPM} \qquad (6.4)$$

where R is the resistor length.

Once again, the ADC-based technique does not require any special devices other than the signal processing device. Therefore, the technique is advantageous as each condition monitoring system is attached with a signal processing device. Hence, the speed estimation from the encoder pulse signal is purely dependent on the signal processing technique.

6.12 Noise Measurements

Microphones are used for noise measurements. For machinery noise measurements, two types of microphones are used. One type of microphone is known as the *capacitor* type or the *condenser* type of microphone. This microphone consists of two metal plates next to each other. The top plate is usually a metal foil of around 10-μm thickness placed above a thick bottom backing plate. These two plates act as a capacitor. A polarization of 200 Volts is usually supplied from a microphone power supply source. When the sound pressure waves are incident on the thin foil, the foil deflects, which corresponds to the change in the capacitance between the two plates. This change of capacitance leads to a charge variation at the output of the microphone. Usually the deflections of the foil are very small and around 0.5 μm. This foil, which acts as a thin diaphragm, is very delicate and is usually protected with a grid-type cover. The microphone plates are circular disks. The capacitor- or condenser-type microphones require a charge-to-voltage amplifier so that pressure signals that are converted to voltage can be transmitted over large cable lengths. The condenser microphones come in sizes of 1/8-inch, ¼-inch, ½-inch, and 1-inch diameters. The smaller the microphone, the greater is its frequency response. For example, a ¼-inch condenser microphone has a flat frequency response up to 70 kHz, and the typical sensitivity of these microphones with a unigain preamplifier is anywhere from 1 mV/Pa to 50 mV/Pa. One of the useful applications of these microphones is that they can withstand high temperatures up to 250°C since they have no delicate electronic component inside the microphone cartridge.

Another type of microphone is the prepolarized type, where no external polarization voltage is required. In the back of the thin foil membrane, piezoelectric material patches are pasted. This piezoelectric material develops a charge when the diaphragm is strained due to an incident pressure wave. The charge-to-voltage conversion is done in a preamplifier attached to the microphone cartridge. These microphones only require a DC supply of 4 mA current for the preamplifier. These types of microphones cannot be subjected to high temperatures due to the presence of the piezoelectric material.

Microphones in industrial applications can be permanently installed for around-the-clock noise monitoring, like in airport noise monitoring, and so on. These microphones are weatherproofed and can withstand high humidity and ambient temperature variations. When microphones are used in conditions with excessive wind, it is a good practice to place a windscreen over the microphone to cut the steady wind noise.

6.12.1 Sound Intensity Measurements

Sound intensity is measured by a two-microphone probe. This probe consists of two phase-matched microphones held together in a small frame with a spacer of fixed length between the two microphones. The two microphones can be either face to face or in a side-by-side configuration. The microphones can be of the condenser or prepolarized type. The accuracy of sound intensity very much depends on the phase matching of the microphones. It is preferred that an in situ phase calibration is followed by a correction in the sound intensity computations. A higher spacer distance is preferred for lower-frequency sound intensity level (SIL) measurement and the spacing decreases for higher-frequency SIL measurements. For a spacing of 12.7 mm for ½-inch microphones, the SIL can be measured to 4000 Hz without any significant errors.

6.13 Temperature Measurements

Temperature is a very important parameter to be monitored, in particular for bearings and shaft couplings. Many instruments are widely used for temperature measurements. At high temperatures, when mounting or access to machines is a problem, noncontact instruments like pyrometers and thermal imaging cameras are appropriate. Such noncontact-based temperature measurements are described in Chapter 10.

Thermocouples and resistance temperature detectors (RTDs) are used for contact-type temperature measurements. In RTDs, the change in the resistance of the material is used to detect temperature. RTDs can be used to measure very high temperature by using platinum wires, which have a high melting point. Usually a Wheatstone bridge circuit is required to detect the change in resistance.

Another very popular temperature measuring device is a thermocouple. When two dissimilar metals are joined together and the two joints are kept at two different temperatures, an electromotive force, or emf, is generated. This emf can be measured by a voltage-measuring device. For different material combinations and temperature differences, standard tables of voltage versus temperature are available. A few common thermocouple material combinations that are used in the industry are the copper-constantan, chromel-constantan, iron-constantan, chromel-alumel, and platinum-platinum 10% rhodium. These thermocouple types are known as the T, E, J, K, and S type thermocouples, respectively. Various thermocouples can be used to measure temperatures from –180°C to 1600°C. Usually, in industrial measurements, one junction of the thermocouple is kept at the location whose temperature is to be measured, and the other junction is maintained at the data acquisition device, which is electrically compensated by a voltage source corresponding to the temperature of the thermocouple at the data acquisition end.

6.14 Laser-Based Measurements

Lasers are being used for measurements in many industrial applications. Lasers are also used for guiding systems in shaft alignment and distance measurements. Some of the applications of lasers in condition monitoring are for measuring the amount of suspended materials in a given volume of liquid and for measuring linear and rotational vibrations of bodies based on the Doppler effect principle.

6.14.1 Laser Vibrometer

In machinery condition monitoring, there are many instances when contacting vibration measurements are not convenient, for example, the temperature of the surface could be very high for the piezoelectric accelerometer to withstand. The surface on which the vibration is to be measured is very small and too lightweight to attach the accelerometer or the surface is not easily accessible. In such cases, it is convenient to have a noncontact-based vibration measurement. Figure 6.8 shows a laser vibrometer being used for refrigerator compressor vibration measurements. Laser-based vibration measurements can be used for single-point transverse measurements, vibration over a two-dimensional surface, or using three laser-based measurement single-point vibrometers and an appropriate geometrical integration software, three-dimensional vibration of a body can also be obtained. Recently, two laser beams separated by a fixed distance are used as a single focusing head to measure rotational or torsional vibration of rotating shafts as well.

Figure 6.9 illustrates the working principle behind a single-point laser vibrometer. Usually, a Ne-He laser source of frequency 4.74×10^{14} Hz is used

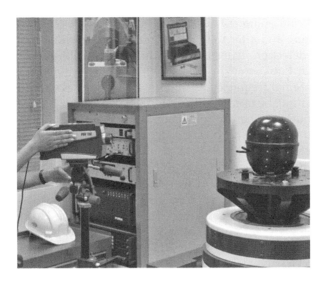

FIGURE 6.8
Compressor vibration measurements using a laser vibrometer.

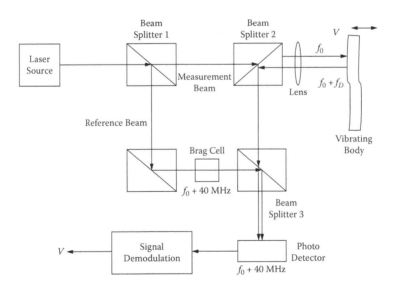

FIGURE 6.9
Block diagram of a laser vibrometer.

in such laser-based vibrometers. At the first beam splitter, the beam is split into a measurement beam and a reference beam. The measurement after passing through the second beam splitter is focused by a lens onto a vibrating surface. Because of the Doppler effect, since the surface is vibrating, the reflected beam with a Doppler frequency of f_d is incident back on the second beam splitter. This shifted beam is made to interfere with the reference beam, and after the third beam splitter, is directed onto a photo-detector for detection. After the demodulator, the vibration velocity is calculated as per Equation (6.5). Since the frequency of the laser beam is very high, conventional frequency estimation techniques like fast Fourier transform (FFT) cannot be used. Rather a signal hetrodyning is done to detect the Doppler frequency. The Bragg cell, an acousto-optic modulator provides a known frequency of 40 MHz, and sidebands on either side of the reference frequency help in the estimation of the direction of the vibration at the measurement point. The intensity of the reflected wave from the surface decides on the signal-to-noise (SNR) of the vibration. In order to receive a strong reflected laser beam, the surface must be shiny or a reflective tape can be put at the measurement point. Usually the lens can be adjusted to focus the beam as a sharp single point on the surface. The noise introduced by improper focusing is known as speckle noise. It is usually advisable to mount the laser vibrometer on a strong tripod, so that unnecessary low-frequency vibrations are not introduced in the vibration measurements. The laser beams can be focused from a distance as high as 30 m from the source whose vibration is to be measured. The vibration velocity thus obtained can be digitally differentiated to obtain vibration in terms of acceleration. In the frequency domain, it essentially means to multiply the vibration velocity by ($j\omega$).

$$f_d = \frac{2V}{\lambda} \qquad\qquad (6.5)$$

Some of the possible engineering applications of laser-based vibration measurements are windmill gearbox vibration monitoring, bridge vibration monitoring, gas turbine health monitoring, and delicate membrane vibrations.

6.14.2 Rotational Laser Vibrometer

A rotational laser vibrometer can be used to measure the instantaneous rotational speeds or torsional vibration of rotating shafts. It consists of two laser sources kept at a fixed distance apart, d, in a single laser head unit as shown in Figure 6.10. A rotational laser vibrometer used to measure torsional vibrations in a rotor rig is shown in Figure 6.11. In Figure 6.10 it is seen that each point on the circumference of a rotating body with angular velocity ω can have a tangential velocity vt allocated to it, which depends on the rotation radius R. This translational velocity can be split into two translational velocity components that are of any orientation, but are at right angles to each other. One of the vector components of the translational velocity is in the direction of the incident laser beam.

As is shown in Equations (6.6), (6.7), and (6.8), by measuring two translational velocity components running in parallel, it is possible to deduce the

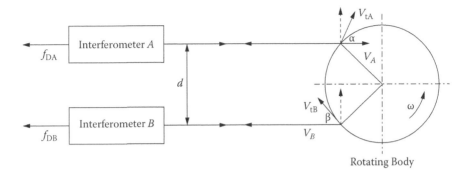

FIGURE 6.10
Acquiring angular velocity through two-point measurement.

FIGURE 6.11
Rotational laser vibrometer being used for torsional vibration measurements.

angular velocity ω. A measurement setup consisting of two interferometers with two laser beams running in parallel at a distance d will acquire the velocity components v_A and v_B. Due to the vector analysis, these two relationships apply:

$$V_A = V_{tA}.\cos\alpha = \omega.R_A.\cos\alpha$$

$$V_B = V_{tB}.\cos\beta = \omega.R_B.\cos\beta$$

(6.6)

The velocity components v_A and v_B, working in the direction of the laser beams, generate the Doppler frequency shifts f_{DA} and f_{DB} in the laser beams scattered back, as explained in the previous section. For this, the following two relationships apply:

$$f_{DA} = \frac{2V_A}{\lambda} = \frac{2(\omega.R_A.\cos\alpha)}{\lambda}$$

$$f_{DB} = \frac{2V_B}{\lambda} = \frac{2(\omega.R_B.\cos\beta)}{\lambda}$$

(6.7)

The geometric connection between the distance d of the laser beams and the angles α and β with the given radii R_A and R_B can be described by:

$$d = R_A\cdot\cos\alpha + R_A\cdot\cos\beta \tag{6.8}$$

From this relationship, the Doppler frequency shifts can be calculated using Equation (6.9).

$$f_D = f_{DA} + f_{DB} = \frac{2d.\omega}{\lambda} \tag{6.9}$$

Thus, the resulting Doppler frequency shift only depends on the system constants d and λ and the angular velocity ω. The contributions of the Doppler content f_{DA} and f_{DB} depend on the relative position of the laser beams to the rotation axis. With a precise symmetrical setup (i.e., rotation axis precisely in the center between the beam axes) the velocity components v_A and v_B and thus also the Doppler content are of equal amount.

6.15 Current Measurements

Current waveforms as a function of time are measured by two methods—the Hall effect method and the magnetic induction effect. Using the Hall effect method, DC current can be measured, whereas for measuring by the induction effect an AC current is required.

6.15.1 Inductive Current Sensor

This current sensor is based on the principle of electromagnetic induction. When an AC current passing through a conductor is surrounded with an electrical coil on a soft core, a voltage is induced in the coil due to electromagnetic induction. This voltage can be used to measure the current in the conductor.

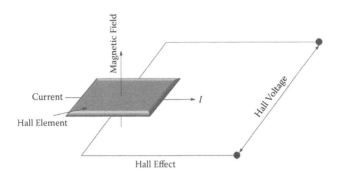

FIGURE 6.12
Hall effect circuit.

This type of sensor does not require any external power source and is easy to install. However, such sensors can only be used to measure AC currents.

6.15.2 Hall Effect Sensor

The magnetic field generated around a current-carrying conductor is sensed by a Hall effect device. A Hall effect device is a four-terminal solid-state semiconductor material commonly composed of materials like silicon, germanium, indium arsenide, or gallium arsenide. When a Hall device is inserted in a soft iron core around the current-carrying conductor and a current is applied in a direction perpendicular to the magnetic field, a Hall effect voltage is generated proportional to the current flowing in the conductor. Essentially, the Hall effect is another way of detecting the magnetic field around a current-carrying conductor. In order to introduce a current in the semiconductor Hall device, an external power source is required. In a Hall effect current sensor, a varying resistor is provided to calibrate the sensor by varying the applied field voltage. The greatest advantage of Hall effect sensors is that they can be used to measure the DC current in addition to the AC waveform. However, the magnitude of current that can be measured with the Hall device is less than that of an inductive current sensor. A Hall effect circuit is shown in Figure 6.12. Figure 6.13 shows a Hall effect current sensor.

6.16 Chemical Composition Measurement

It is often necessary to measure of the amount of chemical elements present in a material. Such a study has wide applications in machinery condition monitoring, for by knowing the amount of a particular element present in a worn particle, a clue to its rate of wear, and in turn the rate of health deterioration of the machine, can be known. The details of such techniques

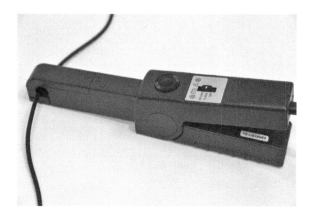

FIGURE 6.13
Handheld Hall effect current probe.

are discussed in Chapter 11. When the electrons are excited in any element, they radiate energy at frequencies that are characteristic of the particular element. This concept is used in the atomic emission and the atomic absorption spectrophotometer.

6.16.1 Atomic Emission Spectrophotometer

In an atomic emission spectrophotometer, the material or chemical compound whose elemental composition is sought, is heated in a small furnace so that the material is ionized; the electrons become excited and jump into an unstable orbit. When they come back to their stable orbit, they radiate energy at frequencies that are characteristic of a chemical element. This emitted radiation is captured by a photodetector and the presence of the constituent chemical elements is known. An atomic emission spectrophotometer (AES) thus is used to get a qualitative feel of the chemical elements present in a material. Since the energy emitted by all the chemical materials are measured simultaneously, the signal-to-noise ratio due to cross talk is poor.

6.16.2 Atomic Absorption Spectrophotometer

An atomic absorption spectrophotometer (AAS) works on a similar principle; however, using an AAS, the chemical elements that are present in a material in percentage composition can be known, one by one. A light source of a frequency that corresponds to a particular chemical element to be estimated is used. When the excited electrons in the outer orbit return to their stable state by releasing energy, only the energy at the frequency that corresponds to the particular chemical element is absorbed. Thus, an estimate of the elemental composition can be made very accurately. AASs are thus used for a quantitative estimation of the chemical elements present in a material. Figure 6.14 shows the block diagram of an AAS, and an AAS in the laboratory is shown in Figure 6.15.

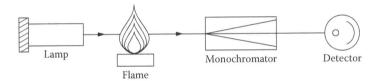

FIGURE 6.14
Block diagram of an AAS.

FIGURE 6.15
View of an atomic absorption spectrophotometer in the testing laboratory.

6.17 Ultrasonic Thickness Measurement

Sound waves of high frequency in the ultrasonic range of 50 kHz to 2 MHz are used for thickness measurement. A piezoelectric probe is used to generate a high-frequency ultrasonic sound wave, which is made to travel inside a material. The wave is reflected back to the source due to an impedance mismatch at the other end of the material. This concept is used to measure the thickness of a material by recording the time it takes for the ultrasonic wave to traverse two lengths corresponding to the thickness of the material, as shown in Figure 6.16. The speed of the ultrasonic wave in the material measures the thickness. For impedance matching of the probe with the top surface of the material whose thickness is to be measured, a water-based gel is usually used as a couplant. This gel allows measurement with a lower signal-to-noise ratio. Such ultrasonic thickness gauges are used for measuring the thickness of gearbox housings, inner wall thickness of pipes, and so on, from the outside. Another application of this principle is used

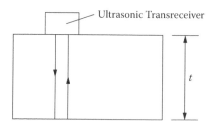

FIGURE 6.16
Ultrasonic wave traversing a sample.

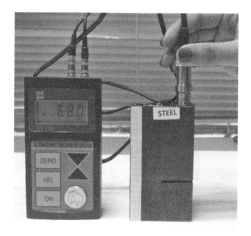

FIGURE 6.17
Ultrasonic thickness gauge.

in ultrasonic flaw detectors, where detection of internal defects in materials that could be due to a casting defect is desired.

The ultrasonic thickness gauges are handheld and thus very portable. They are very convenient for field use and usually have a digital readout showing the thickness of the material. One such ultrasonic thickness gauge being used to measure the thickness of a steel block till a cut location is shown in Figure 6.17.

6.18 Data Recorders

In condition monitoring, in order to archive data and perform analysis on the data acquired from transducers, many times the data needs to be stored for later use. Traditionally, analog recorders were popular, where the signal was recorded in a magnetic media. These recorders worked on a direct recording mode through a magnetization process in the magnetic media, which were usually ferrite-coated plastic tapes. There were improvements in the recording process by performing frequency modulation of the signal, and thus very low frequencies could be recorded. Later, with the development of digital techniques, the tapes were replaced by digital audio tapes (DATs) and recording was done through a digitization process of the FM signals known as pulse code modulation (PCM). Such DAT recorders have a better dynamic range than the FM tape recorders. However, due to the obsolescence of the tape recording media, the use of DAT recorders has decreased.

At present, digital data recorders are more prevalent and are used for field data recording. Such recorders have become convenient since they can store the data on digital media like a memory card. The stored digital data can be very conveniently transferred and shared with computers over USB drives and the Internet. The digital data recorders are multichannel and have modular units for analog data acquisition from various types of transducers, like ICP accelerometers or microphones, thermocouples, strain gauges, optical encoders, and voltage probes. Some of the input modules in such recorders have built-in anti-aliasing low-pass analog filters before the ADC conversion. Such recorders have provision to vary the sampling rate and the amount of data that can be stored. A typical 16-channel, 16-bit data recorder is shown in Figure 6.18. Since the data is stored in the

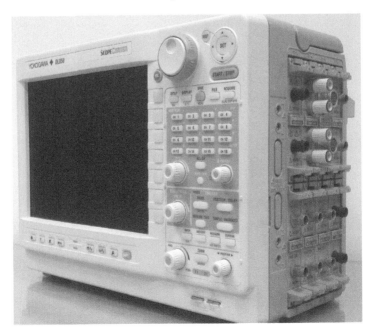

FIGURE 6.18
Digital data recorder.

digital format, simple math operations and even FFT can be done on the data and displayed on the screen. There is a provision to acquire the data whenever an event triggers the data acquisition unit. For field use, such data recorders have the option to operate with a DC power supply like a 12V car battery as well.

7

Vibration Monitoring

In a majority of the rotating machinery, vibration monitoring is preferred. This is due to the fact that every dynamic machine component will manifest itself in the measured vibration response of the machine at its characteristic frequencies. Thus, for fault diagnosis, this provides an important and easy methodology to detect faults in operating machines by signal analysis of the measured vibrations from the machines. Machines consist of rotating shafts supported on bearings, perhaps carrying a gear or a pulley, and connected to another similar machine through a mechanical coupling. The general configuration of a machine in any industry consists of a prime mover, usually an electric motor or sometimes an internal combustion engine, driving a mechanical unit like a fan, blower, pump, compressor, or a gearbox. Such a general configuration is shown in Figure 7.1. In this chapter, we will focus on the basics of vibration monitoring of such generic machines, and then discuss the telltale signs of faults in the vibration spectra.

7.1 Principles of Vibration Monitoring

In order to detect faults in machines by vibration monitoring, the vibration from the machine needs to be measured and then the vibration signal processed to obtain, meaningful information regarding the machine's health condition. The vibrations should be measured close to the bearings supporting the rotating shafts. Wherever possible, the vibrations at any location should be measured in three mutually perpendicular directions. Simultaneous measurement of the vibration in three directions can be done using a triaxial accelerometer. It is always desirable to record the rotating speed of the shafts at the instant that vibration measurements are done, because all the predominant frequencies in the vibration spectra are related to the rotating speeds of the shaft. An in situ field calibration of all the transducers is recommended. The machine conditions and parameters, like its rated power, load conditions, special features, and foundation type should also be recorded. With a handheld portable fast Fourier transform (FFT) analyzer, the vibration spectra can be obtained at the machine site. For diagnostic purposes, when a few more signal processing techniques need to be applied on the vibration signal, it is recommended that the vibration data be recorded in a suitable recorder along with a sketch of the

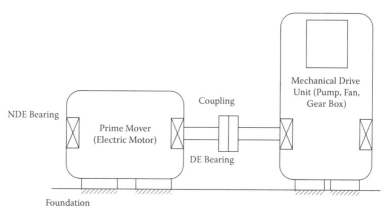

FIGURE 7.1
General machine configuration.

transducer location and its sensitivity, and type. While recording vibration data in the field, it is a good practice to also record the calibration signal as applied to the transducers, so that later on when the data is analyzed in a computer after an analog-to-digital conversion, the user gets an idea of the amplitude of the mechanical parameter on which the corresponding signal voltage level is obtained.

Today, with automated systems in place in many industries, there are very few instances when a maintenance crew actually goes to a machine, installs transducers, and records the vibration. In online automated vibration monitoring systems, the acquisition and analysis of the vibration signal is done through computer software that is resident on a central server having a large database with local terminals all around the plant and a user-friendly graphical user interface. Nowadays, these software systems are even accessible through the appropriate application software in one's mobile phone. When one requires past or present data on a particular machine that has been included in the automated online monitoring system, it can be obtained from any location using such software. Many such commercial software systems are available for online vibration monitoring of machines. This software is developed based on well-established diagnostic routines for some of the commonly occurring faults in rotating machines, some of which are discussed in the subsequent sections.

7.2 Misalignment Detection

Usually it is required that the shafts of the prime mover supported on bearings and the shaft of the driven unit supported on bearings all are in a straight line with the same elevation from the foundation level. The two shafts are joined through a mechanical coupling. However, due to uneven foundation level, the shafts are not concentric and they are offset either by the same lateral amount or by an angle. This condition leads to the misalignment of the shaft, which can be parallel and angular as shown in Figure 7.2. Due to this condition of the shafts, there is an axial push and pull on the shafts and a strong axial vibration is recorded. However, during machine installation, necessary shaft alignments are done by introducing metal shims or adjusting jack bolts at the machine foundation to minimize shaft misalignments.

The procedure for alignment of shafts is given in Appendix A4. Many times, flexible couplings are used to accommodate small misalignments. However, with time, the misalignment may increase because of corrosion of the shims, uneven thermal expansion between the driven and driving ends, and so on. This misalignment leads to an increase in the frictional forces within the flexural elements of the coupling, which also leads to an increase in the temperature of the couplings.

However, through vibration monitoring, the misalignments in shafts can be detected relatively easily because of the increase in the axial vibration levels at a frequency twice the rotating speeds of the shaft. This is because an increase in the misalignment can cause the torque on the shaft to increase at twice the rotational speed. During misalignment, the phase difference at bearings across the coupling are 180°. Misalignment usually provides an elliptical orbit due to a strong force in one direction, as shown in Figure 7.3. In the orbit plot shown in Figure 7.3, the shaft was supported on two rolling-element bearings. Two accelerometers were mounted on the drive-end (DE) bearing housing in mutually perpendicular directions to measure the radial vibrations, as shown in Figure 7.4, which were used to obtain the orbit plot shown in Figure 7.3. Proximity-type eddy current probes are often used to measure the displacements of the shaft ends at a bearing location. The advantage of such probes is that they are noncontacting in nature, and the dynamics of the bearing housing do not influence the measured displacement values.

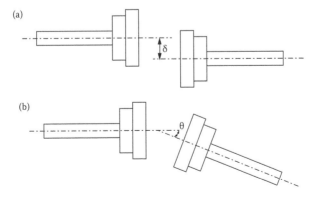

FIGURE 7.2
Shaft misalignment: (a) parallel, (b) angular.

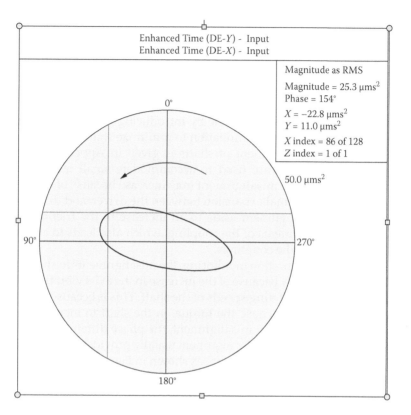

FIGURE 7.3
Orbit plot indicating misalignment in a shafting system.

FIGURE 7.4
Radial accelerometers for orbit analysis.

7.3 Eccentricity Detection

Eccentric rotors occur when the geometric center of the rotor does not coincide with the center line through the support bearings. Eccentricity has serious consequences in electrical machines; because of the eccentricity, the air gap between the rotor and the stator coil varies in every rotation, which gives rise to a varying magnetomotive force and modulates the current drawn by the motor. Similarly, in mechanical rotating machines, eccentricity creates an increase in the radial vibrations at twice the rotational speeds.

7.4 Cracked Shaft

Cracks develop in shafts for many reasons, one of which can be internal defects during manufacturing. Such internal cracks can be detected by ultrasonic testing as is done in the case of rolling stocks. However, surface cracks on the shaft grow with time due to fatigue loading. The crack growth rate is determined by the famous Paris equation, which depends on the material type and type of loading. A schematic of the instrumentation that is required for vibration analysis for fault diagnosis in a rotor bearing system is shown in Figure 7.5. Surface cracks on shafts can be determined by their characteristic vibration spectrum, as shown in Figure 7.6. The characteristic spectrum of a cracked shaft is evident, where the fundamental rotating

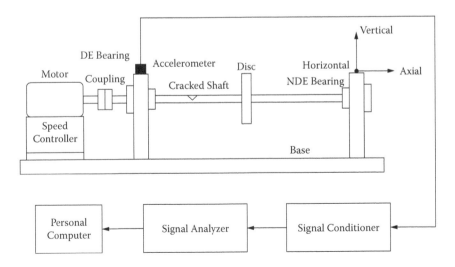

FIGURE 7.5
Instrumentation for vibration monitoring.

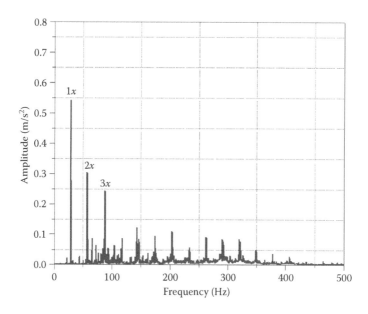

FIGURE 7.6
Vibration spectrum in vertical direction at NDE bearing for rotor at 1740 RPM.

speed along with its first and second harmonics is present. In the case of a cracked shaft, the vibration amplitude reduces by $(2n-1)$, where n is 2 for the first harmonic, and so on.

7.5 Bowed and Bent Shaft

Bowed and bent shafts can be due to manufacturing or installation defects. This machine fault condition exhibits high axial vibration; however, the relative phase difference between the two ends is usually 180°. This phase difference between the two ends is measured by computing the cross-spectrum phase of two transducers in the same direction across the two ends of the shaft. The two ends of the shaft could be the two support bearings of the shaft.

7.6 Unbalanced Shaft

Unbalance is one of the most common types of machinery faults that occurs in rotating machines. Unbalance is due to a net uneven distribution of the mass of a rotating component about its axis. The unbalance force acts radially and is proportional to the square of the angular rotational speed of the shaft. For a typical rotor disc with an unbalance and rotating at 1740 RPM, the vibration spectrum of the vibration signal measured at the nondrive end (NDE) bearing is shown in Figure 7.7.

The unbalance force induces additional fatigue load on the bearings and may lead to premature failure of the bearings and rotating shafts. High-speed machines are precision balanced so that the reaction forces at the bearings are reduced. For long rotors, like those in a multistage compressor or turbine, the unbalance forces can be in different axial planes. Unbalance forces in multiple planes leads to couples as well as radial forces in different planes. Balancing of such long rotors is usually done in multiple planes and is known as *dynamic balancing*. For radial fans, a single-stage pump with one set of impellers, balancing in one plane, suffices. The in situ field balancing techniques for balancing of a disc in a single plane is known as *static balancing* and a couple of methods of balancing are given in Appendix A3. Since the balancing forces are radial in direction, the horizontal and vertical vibrations in the radial plane of the rotating disc are usually the same. The unbalance mass m is at a distance of e from the center of the axis about which the disc is rotating. The grade of the balancing quality is designated as the product of the eccentricity e and the rotational velocity ω in radian/s. Table 7.1 shows the quality of balance grade as per the ISO 1940 standard. The most superior balance quality is G 0.4 and the most inferior

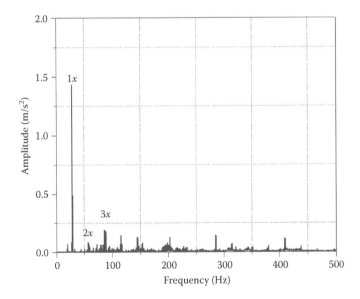

FIGURE 7.7
Vibration spectrum in vertical direction of NDE bearing for unbalanced rotor system at 1740 RPM.

TABLE 7.1

ISO Balance Grade Quality

Quality Grade	ω_e (mm/s)	Application
G4000	4000	Crank shafts of rigidly mounted slow marine diesel engines with uneven number of cylinders
G1600	1600	Crank shaft drives of rigidly mounted large two-cycle engines
G630	630	Crank shaft drives of rigidly mounted large four-cycle and elastically mounted marine diesel engines
G250	250	Crank shaft drives of rigidly mounted fast four-cylinder engines
G100	100	Crank shaft drives of fast diesel engines with six or more cylinders; engines for cars, trucks, and locomotives
G40	40	Car wheels, wheel rims, crank shaft of elastically mounted fast four-cycle engines with six or more cylinders
G16	16	Drive shaft with special requirements as in cars, trucks, locomotives, gasoline and diesel engines, crushing machines, agricultural machinery
G6.3	6.3	Parts of processing-plant machines, turbine gears, central, impellers, and armatures
G2.5	2.5	Gas and steam turbines, turbo generators, compressors, small electric armature
G1	1	Tape recorders, phonographs, grinding machine, small armatures with special requirements
G0.4	0.4	Spindles, discs, and armatures of precision grinders and gyroscopes

balance quality is G4000. Depending upon the application, the balancing grade is selected.

In industry, unbalance can be due to poor manufacturing; for example, in large discs, the presence of casting defects can lead to an uneven distribution of the rotating mass and lead to unbalance. Many times, in chemical processing plants, the chemical slurry gets stuck to the rotating agitator, stirrer, or fan, and if neglected, leads to unbalance and to shearing of the fan blades or damage of the bearings.

7.7 Looseness

In machinery, many components are attached to a rotating shaft. For example, it could be gears and pulleys attached to a rotating shaft or bearings supporting a rotating shaft. There can be instances when these components become loose. Thus, when in motion, they are free to move about and hit one another. This is like significant hammering on the system or like a series of impacts on the rotating machines. For this reason, looseness manifests as high levels of impulsive vibrations in the time domain. The typical frequency domain characteristics of looseness in rotating machines is the presence of vibration peaks at fractional harmonics of the rotational speed and their harmonics.

7.8 Rub

Rub is a phenomenon that occurs in closely placed rotating components against a stationary casing, for example, two closely placed sets of compressor or turbine blades in an aircraft gas turbine engine against the engine casing. When the tips of the blades touch the casing and are rotating at high speeds, they provide a shearing force to the casing. This force is responsible for exciting the structure and leads to its fatigue failure. Since there are many forces occurring due to the rub, the radial forces and vibrations at the bearing housing are at all sub- and super-harmonics of the rotating speed of the shaft.

7.9 Bearing Defects

Bearings are an important machine element that supports the rotor system. They are designed to provide less friction at the supports and carry the loads. In machineries, the two most common types of bearings are used—the antifriction bearing or the rolling element bearing, and the journal bearing working on the principle of hydrodynamic lubrication. Here we will focus on fault diagnosis in rolling element bearings.

Rolling element bearings have four very important components, two circular races, between which the rolling elements—rollers, tapered rollers, or spherical balls—are held. The fourth important element in a rolling element bearing is the cage or the separator. The cage ensures that no two rolling elements come in contact with each other. These elements, usually made of hardened steel, are to be manufactured as perfectly round or spherical surfaces with a very high surface finish. However, due to manufacturing imperfections of grinding, honing, and polishing, these components are noncircular (out of roundness), though the radial variations could be only few microns. Thus, the bearing race surfaces are wavy. In addition, the surface finish may not be adequate and may have a surface roughness, which can be characterized by a center line average value. So, when the bearing elements are in motion, a vibration wave is generated in the bearing. Thus, because of waviness and surface roughness, even new bearings give rise to vibrations, though there are ways by which the vibration can be reduced, for example, by having better manufacturing tolerances and lubrication between the elements.

A representative sketch of the two most important manufacturing imperfections of waviness and surface roughness of a bearing race is shown in Figure 7.8. The radial imperfections of some of the best bearings manufactured in the world are around 3 to 5 μm.

The bearing vibration is further complicated when the radial loads on the bearing change during rotation of the shaft, and defects like impurities or scratches, pits, and so on, develop on the bearing element surfaces. The vibration of the bearing is thus amplitude modulated due to varying load at the frequency of rotation. To obtain the characteristic vibration signals, first the bearing vibration signals have to be demodulated, and then a frequency domain analysis done. Many commercial software has provisions to perform demodulation of the signals. Such signal demodulation is also known as *envelope analysis*, as was discussed in Chapter 5.

The characteristic bearing defect frequencies for the different rolling elements of bearings can be determined as discussed in the following paragraphs.

With reference to Figure 7.9, for a rolling element (ball) bearing, let the linear velocity of each ball center, V_c, be $V_c = \dfrac{V_i + V_o}{2}$; thus the rotational velocity of the center of the ball is given by Equation (7.1).

$$\omega_c = \frac{1}{2}\left[\omega_i\left(1 - \frac{BD\cos\phi}{PD}\right) + \omega_0\left(1 + \frac{BD\cos\phi}{PD}\right)\right] \tag{7.1}$$

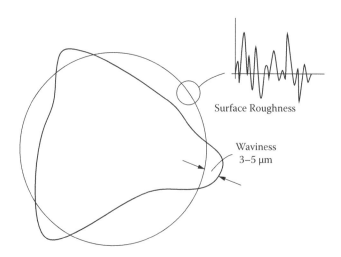

Surface Roughness

Waviness
3–5 μm

FIGURE 7.8
Schematic representation of waviness and surface roughness in a bearing race.

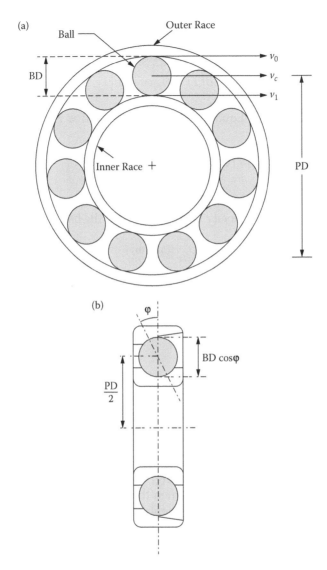

FIGURE 7.9
Geometry of a rolling element bearing (a) front view, (b) side view.

The ball pass outer frequency is given in Equation (7.2).

$$BPFO = N_b \left| \omega_c - \omega_o \right| \tag{7.2}$$

By using Equation (7.1), Equation (7.2) can be written as Equation (7.3).

$$BPFO = \left| \frac{Nb}{2} (\omega_c - \omega_o) \left(1 - \frac{BD \cos \phi}{PD} \right) \right| \tag{7.3}$$

Similarly, the ball pass frequency inner (BPFI) can be described as the number of balls multiplied by the different frequency between the inner race and the cage, and is given as Equation (7.4).

$$BPFI = N_b \left| \omega_i - \omega_c \right| \tag{7.4}$$

By using Equation (7.1), Equation (7.4) can be written as Equation (7.5).

$$BPFI = \left| \frac{Nb}{2} (\omega_i - \omega_c) \left(1 + \frac{BD\cos\phi}{PD} \right) \right| \tag{7.5}$$

Considering the linear velocity of a point on the inner race in contact with the ball surface, the linear velocity (V_b) of a point on the ball surface is given as

$$V_b = \frac{(\omega_i - \omega_c)}{r_i} \tag{7.6}$$

The ball spin frequency (BSF) is then given as

$$BSF = \left| (\omega_i - \omega_c) \frac{r_i}{r_b} \right| \tag{7.7}$$

Again, using Equation (7.1), the BSF can be written as

$$BSF = \left| \frac{PD}{2BD} (\omega_i - \omega_o) \left(1 + \frac{BD^2 \cos^2\phi}{PD^2} \right) \right| \tag{7.8}$$

Equations (7.1), (7.3), (7.5), and (7.8) are the general equations where either bearing, race, or both, could be rotating.

For the usual case in a machine when the outer race is fixed and the inner race is rotating at N rpm, the fundamental bearing frequencies are given by

$$FTF = \frac{N}{120} \left(1 - \frac{BD\cos\phi}{PD} \right) \tag{7.9}$$

$$BPFO = \frac{N}{120} N_b \left(1 - \frac{BD\cos\phi}{PD} \right) \tag{7.10}$$

$$BPFI = \frac{N}{120} N_b \left(1 + \frac{BD\cos\phi}{PD} \right) \tag{7.11}$$

$$BSF = \frac{N}{120} \frac{PD}{BD} \left(1 + \frac{BD^2 \cos^2\phi}{PD^2} \right) \tag{7.12}$$

where FTF is the fundamental train frequency.

In this section, the vibration spectra of measured vibration signals from bearings with seeded faults in a rotor-test rig operating at 1740 RPM are shown. The bearings were ball bearings of 33.5-mm pitch diameter having 8 balls of 7.94-mm ball diameter. Figure 7.10 shows the vibration spectrum at the NDE bearing in the vertical direction when a normal bearing was supporting a shaft rotating at 1740 RPM. The vibration spectrum was obtained by amplitude demodulating the measured vibration signal followed by a frequency analysis. Figures 7.11 to 7.14 show the vibration spectra for bearings seeded with artificial faults in the inner race, outer race, ball, and all bearing components defective, respectively. The defects in the bearings were introduced by having a single groove of 1-mm width etched along the entire width of the bearing seat on the respective bearing surfaces like the inner race, outer race and ball. The defects in the bearings are considered as surface defects.

Figures 7.11 to 7.14 show the low-frequency vibration of the defective bearings, where the higher harmonics of the characteristic defect frequencies of the bearing components are present. However, there is another development in the high-frequency spectrum of the defective bearings. This is explained as follows. The bearing components have a natural frequency due to their mass and stiffness and usually are designed so that their natural frequency is beyond the operating frequencies of the machine, so that any resonance condition is avoided. However, incipient faults develop in bearings due to formation of pits in the races, which give rise to impact excitation of the races when the rolling elements go over such pits. These impact forces, which are

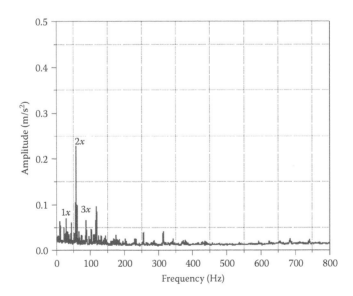

FIGURE 7.10
Demodulated vibration spectrum in vertical direction of normal NDE bearing at 1740 RPM.

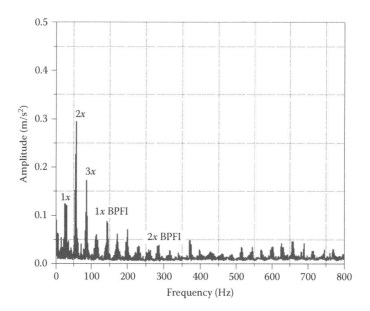

FIGURE 7.11
Demodulated vibration spectrum in vertical direction of inner race defect NDE bearing at 1740 RPM.

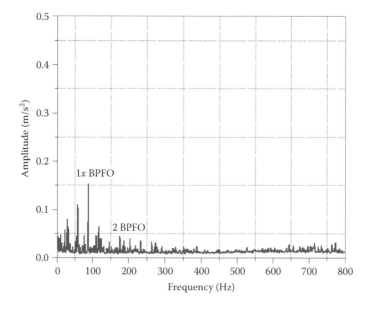

FIGURE 7.12
Demodulated vibration spectrum in vertical direction of outer race NDE bearing at 1740 RPM.

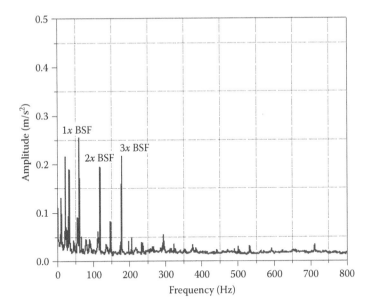

FIGURE 7.13
Demodulated vibration spectrum in vertical direction of ball defect NDE bearing at 1740 RPM.

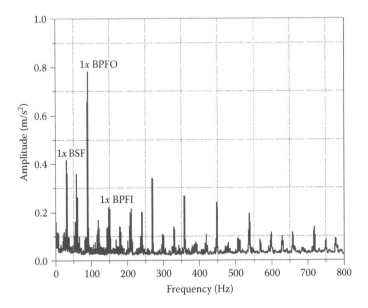

FIGURE 7.14
Demodulated vibration spectrum in vertical direction of all-component defect NDE bearing at 1740 RPM.

of short time duration are in fact responsible for high-frequency excitation of the bearing races. These high-frequency excitations are responsible for the resonant vibration of the races, which thus produces a high level of vibration at high frequencies. In typical rolling element bearings, these vibrations are at frequencies beyond 20 kHz. So by detecting the presence of high-frequency vibrations, bearing defects can be easily detected. Some commercial bearing fault detectors usually monitor these high-frequency vibrations only to detect a bearing defect.

7.10 Gear Fault

Gears are responsible for power transmission between two shafts at the desired rotation speeds. Gears are of different configurations where the power transmission shafts can be either parallel or perpendicular to each other. However, during meshing, the gears are in contact with each other, and as soon as they leave contact due to a sudden decrease in the torque transmission, the gears are set into torsional vibration, which is transmitted to the shafts, bearings, and finally the gear box housing or casing. Gearboxes are a very important machinery in any plant. Many operations require that the shaft rotate at slow speeds. The prime mover, usually an electric motor, rotates at speeds slightly less than its synchronous speed. For example, a four-pole electric motor with a supply frequency of 50 Hz rotates at 1440 RPM. However, when the process operation requires a slow speed of, say, 5 RPM, a large speed reduction has to be brought about by a multistage speed-reducing gearbox. In such gearboxes, due to the heavy loads, a lot of heat is generated while the gears mesh; thus, they are always lubricated to reduce the friction and transfer the heat generated at the gear-meshing location.

There are many ways by which such defects in gearboxes can be detected, one of which is to monitor the lubrication condition and the nature and type of gear wear debris deposited in the lubricating oil. Such a method of fault detection will be discussed in a later chapter.

However, by vibration measurements, gear faults can be detected through appropriate signal processing of the measured vibration signals. The characteristic of a gear vibration is a modulated signal that could be amplitude or frequency modulated. These modulated vibration signals give rise to sidebands around the gear meshing frequency in the vibration spectrum. Thus, the characteristic of a gear vibration signal is the presence of a family of sidebands around the gear meshing frequency and its harmonics. However, in conventional spectral analysis, due to the presence of bearing characteristic frequencies and other unwanted machinery vibration, many times it becomes difficult to identify this family of sidebands.

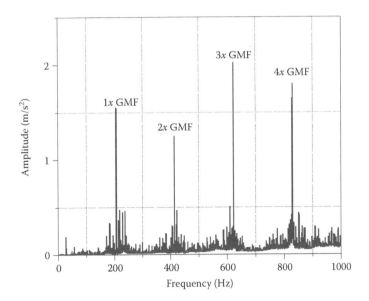

FIGURE 7.15
Vibration spectrum in vertical direction on normal gearbox for pinion shaft at 11.6 HGz.

For example, the vibration spectrum measured on a normal gearbox with the pinion rotating at 1740 RPM (29 Hz) under a load of 0.45 N-m is shown in Figure 7.15. In order to identify a family of sidebands, cepstrum analysis is a well-established technique and has been used here to analyze the vibration signal of the normal gearbox at 1740 RPM. In Figure 7.16, the frequency of 4.8 ms corresponds to the gear meshing frequency of 208.3 Hz while the pinion is rotating at (29/2.5 = 11.6 Hz). The pulley that was driving the gearbox pinion with power from the rotor shaft had a diameter ratio of 2.5, as shown in Figure 7.17. In a defective gearbox, the amplitudes at the corresponding frequency would increase.

For spur gears, the gear mesh frequency is the product of the number of gear teeth and the rotational speed as given in Equation (7.13)

$$GMF[\text{Hz}] = \frac{N_G T_G}{60} = \frac{N_P T_P}{60} \tag{7.13}$$

where N_P and N_G are the rotational speed of the pinion and gear in RPM, and T_P and T_G are the number of teeth in the pinion and gear, respectively.

7.10.1 Gear Meshing Frequency for Planetary Gear

Many times, planetary gear sets are used in machines where multiple output speeds can be obtained. The configuration of a planetary gear with the

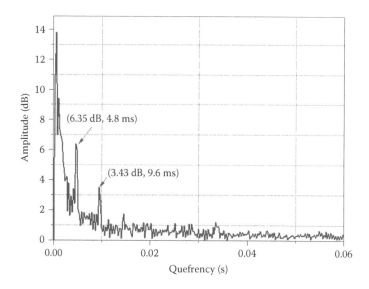

FIGURE 7.16
Cepstrum of vibration signal from normal gearbox with pinion at 11.6 Hz.

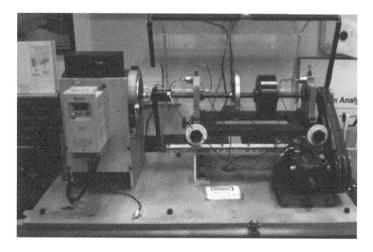

FIGURE 7.17
View of the gearbox being driven by a belt drive in a rotor rig.

sun gear in the center surrounded by a ring gear and the planet gears in between is shown in Figure 7.18. The speeds of the sun (N_s), planet (N_p), and ring (N_r) gears and the carrier (N_s) and the sun–planet gear mesh ($GM_{s/p}$) and planet–ring gear mesh ($GM_{p/r}$) are given in Equations (7.14) to (7.19). Speeds of two of the rotating components and the number of teeth on all

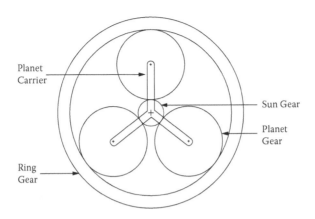

FIGURE 7.18
A representative planetary gear.

gears must be known to calculate the speeds of the other two components and gear mesh frequencies.

$$N_s = N_r \left(\frac{T_r}{T_s} \right) + N_c \left(1 - \frac{T_r}{T_s} \right) \qquad (7.14)$$

$$N_c = \frac{N_r \left(\frac{T_r}{T_s} \right) - N_s}{\left(\frac{T_r}{T_s} \right) - 1} \qquad (7.15)$$

$$N_r = N_s \left(\frac{T_s}{T_r} \right) + N_c \left(1 - \frac{T_s}{T_r} \right) \qquad (7.16)$$

$$N_p = N_r \left(\frac{T_r}{T_p} \right) + N_c \left(1 - \frac{T_r}{T_p} \right) \qquad (7.17)$$

$$GM_{s/p} = |(N_s - N_c)|T_s \qquad (7.18)$$

$$GM_{s/p} = |(N_r - N_c)|T_r \qquad (7.19)$$

7.11 Faults in Fluid Machines

Fluid machines are energy-conversion devices in which mechanical work is either produced (pump) or absorbed (pump) due to the momentum change of a stream of fluid. Fluid machines can be broadly classified into two categories—the rotodynamic machine and the positive displacement machine. The rotodynamic machine, as the name suggests, is primarily rotary in nature and the fluid flow is continuous. In a positive displacement machine, the fluid flow is not continuous and a quantity of fluid is always moved together as a single unit. In a given space, the inlet and outlet ports do not open at the same time; hence, high pressure differentials can be obtained. Since the fluid is bound on all sides, the volumetric capacity of positive displacement machines is limited in comparison to rotodynamic machines. Reciprocating pumps and compressors, vane pumps, gear pumps, and compressed air motors are examples of positive displacement machines. Turbines, centrifugal and axial flow pumps, fans, blowers, and rotary compressors belong to the category of rotodynamic machines. The set of blades in a pump are known as its *impeller*, whereas in a turbine this is known as its *runner*. A fan, a blower, or a compressor belong to the same family of fluid machines handling mostly air, and a pump refers to the fluid machines that handle liquid.

The defects in these machines lead to loss of performance, which can be detected in the process parameters like upstream–downstream pressure and flow rate. There are leaks across the seals between the machine casing and the rotating shaft in multistage compressors. However, defects due to bearing faults, blade unbalance, missing blades, and blade cracks can be detected by conventional vibration monitoring. The vibration spectra obtained from such fluid machines usually has a strong presence of the vane pass frequency or the blade pass frequency, which is a product of the rotational speed of the rotor and the number of blades and vanes attached to the rotating shafts. These vibration signals are modulated, and thus in the case of severe defects there is a strong presence of sidebands around the vane pass or blade pass frequency.

7.11.1 Cavitation-Induced Vibration

In fluid machines at high flow rates in the delivery end, the pressure at the suction end can be lower than the atmospheric pressure, which could reach the limit of the fluid vapor pressure, and sporadic vapor bubbles are formed. Due to buoyancy, the bubbles formed in the low fluid-pressure region move toward high-pressure locations. The difference of pressure between the fluid and the interior of the bubble consequently increases. At some locations, the force of differential pressure exceeds the force of vapor–liquid surface tension and the bubble bursts, creating a high-frequency vibration. The phenomenon of formation and collapsing of bubbles in such fluid machines is known as *cavitation*. Due to cavitation, the internal walls of the inlet pipelines become fatigued and damaged over time. Thus, the presence of high-frequency random vibrations in fluid machines is an indication of the presence of cavitation and subsequent damage of the fluid machine.

8

Noise Monitoring

8.1 Introduction

When a machine develops a defect while in operation, the result is usually excessive noise and vibration, though by performing a detailed signal analysis of the vibration of the machine the defect along with its cause can be determined. However, on the shop floor where the machine is located, one's attention is drawn toward the machine because of the noise produced by it. Thus an understanding of the basics of noise is required, which is presented in this chapter.

Acoustics is the study of the generation, propagation, and reception of sound that is heard by a human being. Sounds heard by human beings are perceived as desirable or undesirable. The undesirable sound is traditionally known as *noise*. There are many sources of noise like machinery noise, traffic noise, animal noise, ambient noise, and so forth. The sound is due to the sound pressure waves incident on the eardrum of the human being. These sound waves require a elastic medium for their propagation, and the nature of these waves can be longitudinal, transverse, bending, shear, or surface. In air, the sound waves propagating are longitudinal in nature and in solids they are usually shear or bending. The speed of propagation of the wave front is dependent on the nature of the wave and the properties of the medium.

The amplitude of the pressure wave that is incident on the eardrum provides a sensation of hearing. Human beings are able to hear sound waves in a particular frequency range only, which is known as the *audible frequency range* of 20 Hz to 20 kHz. Any sound wave with a frequency below 20 kHz is known as *infrasonic* and sound waves with a frequency above 20 kHz are known as *ultrasonic*.

8.2 Acoustical Terminology

While dealing with noise monitoring, there are certain acoustical terms that need to be understood.

8.2.1 Sound Pressure Level

The sound heard is basically due to time-varying pressure waves incident on the eardrum. These time-varying pressure waves are small perturbations of pressure waves above the atmospheric pressure of the air incident on the human ear generated due to a noise source. For example, if at standard temperature conditions of 20°C at mean sea level, the atmospheric pressure at a place is of the order of 10^5 pascals, a small pressure perturbation of even 1 Pa over the atmospheric pressure is loud enough to be heard by human beings. Human beings can hear sound pressures as low as 20 μPa to as high as 200 Pa. Thus, to represent such a large variation in the sound pressure amplitude levels, a logarithmic scale is preferred over a linear scale. The sound pressure level (SPL) is calculated as per Equation (8.1) and expressed in decibels (dB).

$$SPL = 20 \log_{10} \left(\frac{p}{p_{ref}} \right) \qquad (8.1)$$

where p is the root mean square amplitude of the pressure wave and p_{ref} is the reference sound pressure. Internationally, while sound pressure in air is being considered, this reference value of the sound pressure is taken as 20 μPa. Thus, a sound pressure of 20 μPa will correspond to an SPL of 0 dB. A sound pressure of 1 Pa corresponds to an SPL level of 94 dB and that of 200 Pa corresponds to an SPL of 140 dB.

The sound pressure level of a machine is dependent on the environment around it and the distance of the measurement location from the machinery. Thus, while reporting the measured SPL of a machine, the environment, background noise, and the distance from the machine needs to be reported apart from the ambient temperature, pressure, humidity, wind speed, and direction.

8.2.2 A-Weighting

Human beings perceive the same level of sound differently at different frequencies, and their response to the sound pressure level incident on the eardrum is not linear. Further, this perception is different at different amplitude levels. To account for this perception of human beings, a correction is usually applied to the measured sound pressure level of a particular loudness level at different frequency bands. When this correction factor is applied to the measured sound pressure level, there is good correlation between human perception and the weighted sound pressure level. This weighted sound level at normal sound pressure levels is known as the *A-weighted sound pressure level*. The sound pressure level is then denoted as dBA. At very high amplitude levels of sound pressure, like aircraft noise measurements, a different weighting scale is used, and the sound pressure then is *C-weighted* and the overall SPL is represented in dBC. An A-weighting actually brings down the SPL at frequencies below 1000 Hz. In almost all industrial measurements, reporting of the A-weighted overall sound pressure level in dBA is preferred over the unweighted SPL.

8.2.3 Sound Power Level

The sound power level, denoted as SWL, is the amount of acoustical energy contained in the system. SWL is also expressed in a dB scale and can be A-weighted as well. The reference sound power level is internationally taken to be 10^{-12} W. Equation (8.2) denotes the sound power level of a source

$$SWL = 10\log_{10}\left(\frac{W}{W_{ref}}\right) \tag{8.2}$$

Since the SWL of a machine or system is inherent to it, this SWL does not change with the environment or weather conditions. Thus, SWL is always preferred to rate the sound-producing capacity of a machine, though machine designers and noise control engineers always try to reduce the SWL of the source to make quieter products.

8.2.4 Sound Intensity Level

Sound intensity is the product of the sound pressure and the particle velocity and can be expressed as the normal sound power radiated over an area. The sound intensity is also denoted using the decibel scale as given in Equation (8.3)

$$SIL = 20\log_{10}\left(\frac{I}{I_{ref}}\right) \tag{8.3}$$

where again, internationally, in air I_{ref} is taken as 10^{-12} W/m². Sound intensity is a vector quantity and thus has a direction associated with it. Sound intensity thus enables one to determine the source of a noise being produced by a machine. The sound intensity technique is also a very convenient method to measure the radiated sound power of a machine. The sound intensity of plane waves and three-dimensional sound waves can be estimated by determining the sound pressure and particle velocity. At large distances from the source, when the product of the wavenumber and the distance from the source is very high, the magnitude of the three-dimensional intensity can be approximated to be equal to that of plane waves.

8.2.5 Octave Frequency Bands

Sound produced by machines contain many frequencies, which are related to the operational dynamics of the machine. For example, in the noise produced by a blower, there could be frequencies related to the rotational speed of the blower and the impeller vane pass frequency present. The impeller vane pass frequency is the product of the rotational shaft speed and the number of vanes in the impeller of the vane. Usually, a frequency analysis of the recorded noise is done to determine these frequencies. As reported earlier, there are several ways by which frequency analysis can be done; a filtering technique is one such method. Filters of different frequency bands can be selected. Traditionally, the human ear is able to notice a difference in the frequencies in the filters if one frequency is double the other. Each such doubling of the frequency is known as an *octave band.*

In an octave band filter, the upper frequency is twice the lower frequency, and the center frequency of the filter is the geometric mean of the lower and upper frequencies, as given in Equations (8.4) and (8.5)

$$f_u = 2.0^{1/1} \times f_1 \qquad (8.4)$$

and the center frequency of the band is given as

$$f_c = (f_1 \times f_u)^{1/2}. \qquad (8.5)$$

For a 1/3 octave band filter, the upper frequency is $2^{1/3}$ times the lower frequency. The entire audible band of 20 Hz to 20,000 Hz can thus be divided into octaves or 1/3 octave bands as shown in Table 8.1. However, with digital frequency analyzers with a fixed linear frequency resolution, much finer frequency resolution can be obtained to identify a particular frequency. The weighting values described in an earlier paragraph are available for each of the octave or 1/3 octave bands.

The noise time history and spectrum for the complete human audio range of 20 Hz to 20 kHz of a single-cylinder reciprocating compressor driven by an electric motor through a belt-and-pulley drive as measured by a microphone at a distance of 1 meter from the compressor is shown in Figure 8.1 (a and b). The same noise signal is presented in the 1/1 octave, 1/3 octave, and 1/12 octave band in Figures 8.2 (a, b, and c). The spectrum has a fixed frequency resolution of 20 Hz for the complete audio range. However, the octave bands in Figure 8.2 have a fixed percentage band distribution of the measured noise signal. The A-weighting has been applied to the same signal and 1/3 octave A-weighted noise spectrum is shown in Figure 8.3. The low-frequency A-weighted bands have a lower sound pressure level compared to the no-weighted or linear-weighted bands. The overall SPL of the compressor is 82 dBA in the A-weighted band and the same measured sound is 90 dB in the linear-weighted band. The A-weighting values for the different octave bands are given in Table 8.1.

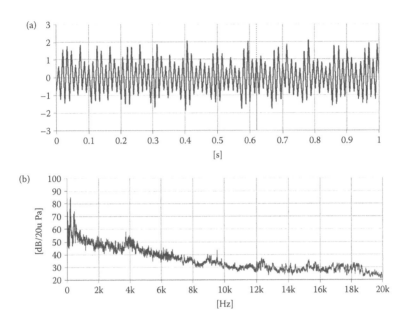

FIGURE 8.1
Measured compressor noise: (a) time signal, (b) linear narrow band spectrum.

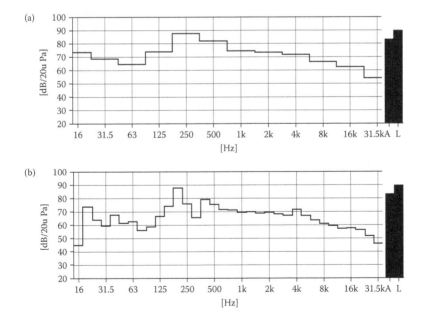

FIGURE 8.2
Linear octave band spectrum of compressor noise: (a) 1/1 octave band, (b) 1/3 octave band.

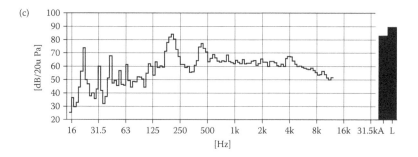

FIGURE 8.2 (*Continued*)
Linear octave band spectrum of compressor noise: (c) 1/12 octave band.

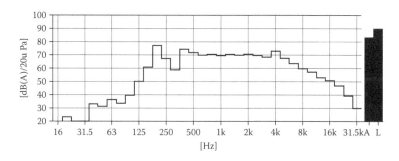

FIGURE 8.3
A-weighted, 1/3 octave band spectrum of compressor noise.

8.3 Noise Sources

The noise sources can be considered to be one-dimensional, two-dimensional, or three-dimensional. Two important parameters are to be considered—the dimension of the body and the frequency of the sound wave. For example, a sound wave can be considered to be one-dimensional or a plane wave if the following condition is satisfied:

$$f < 0.586 \frac{c}{d} \tag{8.6}$$

where c is the speed of sound in air and d is the internal diameter of the circular tube in which the sound wave front is propagating.

When the dimensions of the body are considerable in all the three directions, we need to consider it as a three-dimensional source. However, at large values of ka the source can be considered as a point source, where k is the wave number and a is the characteristic dimension of the body.

TABLE 8.1

Octave and 1/3 Octave Frequency Band Limits along with the A-Weightings

Band Number	1/1 Octave Bands			1/3 Octave Bands			A-Weighting Correction (dB)
	f_l	f_c	f_u	f_l	f_c	f_u	
12	11	**16**	22	14.1	16	17.8	
13				17.8	20	22.4	
14				22.4	25	28.2	
15	22	**31.5**	44	28.2	**31.5**	35.5	−39.4
16				35.5	40	44.7	
17				44.7	50	56.2	
18	44	**63**	88	56.2	**63**	70.8	−26.2
19				70.8	80	89.1	
20				89.1	100	112	
21	88	**125**	177	112	**125**	141	−16.1
21				141	160	178	
23				178	200	224	
24	177	**250**	355	224	**250**	282	−8.6
25				282	315	355	
26				355	400	447	
27	355	**500**	710	447	**500**	562	−3.2
28				562	630	708	
29				708	800	891	
30	710	**1000**	1420	891	**1000**	1122	0
31				1122	1250	1413	
32				1413	1600	1778	
33	1420	**2000**	1840	1778	**2000**	2239	+1.2
34				2239	2500	2818	
35				2818	3150	3548	
36	2840	**4000**	5680	3548	**4000**	4467	+1.0
37				4467	5000	5623	
38				5623	6300	7079	
39	5680	**8000**	11,360	7079	**8000**	8913	−1.1
40				8913	10,000	11,220	
41				11,220	12,500	14,130	
42	11,360	**16,000**	22,720	14,130	**16,000**	17,780	
43				17,780	20,000	22,390	

8.4 Sound Fields

The sound waves that emanate from a machinery source travel in all directions. However, depending on the interaction at the boundaries, they are absorbed, or reflected, or even transmitted further, or a combination of these phenomena occurs depending on the nature of the boundary. This is due to the impedance of the boundary. The sound waves are thus amplified or attenuated depending upon the boundary, and these conditions lead to different sound fields.

8.4.1 Near-Field Condition

The region very close to the source of sound has a sound level that is almost constant and equal to the maximum level of the generated sound. Thus, while monitoring the noise of a machine, measurements in this region should be avoided. The approximate distance from the machine where such a near-field condition exists is equal to one length of the largest dimension of the machine.

8.4.2 Far-Field Condition

Any region beyond the near field of the machinery is known as its *far field*. The far field can be further divided into two distinct regions—the free field region and the reverberant field region. In the free-field region it is assumed that there are no strong reflected waves, and for a point spherical source, the sound pressure level reduces by 6 dB for every doubling of the distance from the source. While performing noise monitoring of machines, it is preferred that the noise is measured in the free-field conditions where there are no reflections from the boundary.

The second region in the far field close to the boundary is known as the *reverberant field* where, because of strong reflections from hard walls, the sound level increases. On a shop floor where machinery noise is to be monitored, measurements away from hard reflecting walls should be done. The free condition of decrease of 6 dB for every doubling of the distance for a three-dimensional point source is violated. Figure 8.4 shows the variation of the noise field from a distance of r from the source.

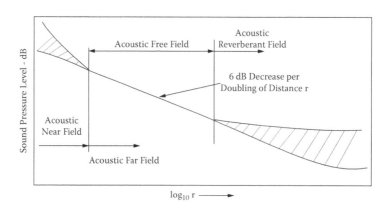

FIGURE 8.4
Variation of sound pressure level at different field conditions.

8.5 Anechoic Chamber

In practice, to obtain free-field conditions, walls are made to be highly absorbing such that no sound waves are reflected back to the source. Special chambers are constructed with all the walls made out of triangular wedges of highly absorbing sound materials. Usually, glass wool or fiberglass-lined wedges are used. The wedges are triangular because if any reflected wave occurs, the geometry of the triangular cross-section wedges causes it to again become incident on the absorbing wedges, and eventually there is hardly any reflection from the walls. Such chambers are used to measure the radiated SPL of a product in free-field conditions. In order to accommodate heavy machinery or vehicles that cannot be placed on a soft fiberglass wedge, a hard floor of concrete is preferred and such chambers are known as *hemi-anechoic* or *semi-anechoic* chambers. Such chambers are very common with vehicle manufacturers who do regular noise measurements and development work on vehicle noise. An essential specification of the anechoic chamber is the frequency beyond which the free-field condition exists, which is known as the *cutoff frequency* of the anechoic chamber, which is essentially related to the dimensions of the chamber. The larger the dimensions, the lower the cutoff frequency of the chamber is. Another essential parameter is the lowest ambient level inside the chamber. Care is taken in the construction of such chambers so that no structure-borne waves affect the noise field. Such chambers have absolute noise levels around 20 dB and follow a particular noise criterion rating.

However, in an industrial shop floor it is very difficult to have a free-field condition or an anechoic chamber for noise monitoring. Thus, noise monitoring is not a popular method for condition monitoring of machines, but it is widely used for ambient noise surveys in shop floors or surroundings at different times of a day. There are standards as to the maximum sound level in a particular environment, and everyone concerned takes care to ensure that such standards and regulations are followed.

8.6 Reverberation Chamber

In contrast with the free-field condition that exists in the anechoic chamber, there are chambers made of hard reflecting walls, where the sound waves have multiple reflections and the sound level is almost the same at every location in the chamber. Such chambers are known as *reverberant chambers*. The sound field is known as diffused in this chamber. Such chambers are used to measure the in situ sound-absorbing coefficient of sound-absorbing material, by measuring a change in the reverberation time of the chamber. A combination of reverberant chamber and anechoic chamber is used to measure the sound transmission loss of panels.

8.7 Noise Measurements

Many times, because of compliance to regulations, the radiated noise levels are measured. There are many standards available for such measurements, or one may develop a protocol of their own for periodic monitoring. For periodic monitoring, the same ambient noise conditions must be ensured as much as possible. Noise measurements are done using an omnidirectional microphone with an adequate frequency range, located about 1 to 1.2 meters from the level ground at a distance of 1 meter from the source as far as possible in free-field conditions. The microphone is usually mounted in a height-adjustable tripod. It is desirable to record the ambient conditions of temperature, humidity, and wind conditions at the measurement site. If the wind speed is high, a windscreen may be put on the microphone to reduce any wind-induced turbulent flow noise. For record keeping and conveying information to the diagnostic engineer, a sketch of the machine, its surroundings, and the measurement location should be done. A field calibration of the microphone both before and after the measurements should be done. Suitable corrections should be applied to the measured values. Handheld microphone field calibrators are available that give 94 dB and 114 dB at 1000 Hz, when the microphone is inserted into the coupler cavity of the calibrator.

Many times, for noise analysis, the sound can be recorded and its frequency analysis done, as was done for the case of compressor in an earlier section.

8.8 Noise Source Identification

For machinery troubleshooting or noise control, it is always desirable to rank and characterize the noise source in terms of its radiated sound pressure level and the frequency content. In a free-field condition when there are no reflected sound waves, the noise sources of a machine can be determined by measuring the SPL at locations on an imaginary conformal volume around the machine. Such measured SPL can be used to produce a contour map of the noise around the machine. Because the sound pressure is a scalar quantity, it doesn't give any information about the direction of the sound, like whether the noise is leaking into a machine or leaking out of a machine. When sound intensity is similarly measured around a machine and the sound intensity vector plotted, the direction of flow of the sound energy can be obtained. Figure 8.5 shows a view of the sound intensity measurement on a portable generator using a handheld two-microphone sound intensity probe.

FIGURE 8.5
Sound intensity measurements on a portable generator.

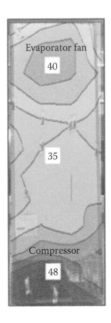

FIGURE 8.6
Sound intensity level contour of the rear of a frost-free domestic refrigerator.

In order to determine the noise sources in a refrigerator, the noise source identification was done using the sound intensity method. The overall SIL contour of the rear side of the domestic refrigerator shown in Figure 8.6 shows the two important noise sources in the refrigerator as the exposed hermetically sealed compressor at the bottom of the refrigerator and the evaporator fan in the top of the refrigerator in the freezer compartment.

9

Electrical Machinery Faults

9.1 Introduction

Electrical motors are used as prime movers in more than 90% of the mechanical drives in industries around the world. The most common mechanical units in industries that are driven by an electrical motors are gear boxes, fans, compressors, and pumps. The fault detection and diagnosis techniques of these mechanical units are usually vibration based and were discussed in earlier chapters. In this chapter, focus is on detection of faults in induction motors as well as the mechanical systems being driven by them by monitoring the stator current of the motor. Discussion of fault detection on nonrotating electrical equipment on power transformers and switch gears using various condition monitoring techniques is also presented.

9.2 Construction of an Electric Motor

Electrical motors have two important elements—the rotor containing coils over a soft core, which is placed inside a set of coils forming the stator held around the frame of the motor body. A balanced power supply, usually a three-phase supply, is provided to the stator, which is responsible for creating a rotating magnetic field; the magnetic field is responsible for the rotation of the rotor. The rotor is sometimes known as the *armature* of the motor and stator the *field winding*. The rotor is supported on bearings at two ends of the motor frame and placed inside the stator of the motor, with a uniform air gap between the stator and the rotor. The frame of the motor has fins around it for dissipation of the heat generated inside the motor while in motion. For large motors, a fan is usually attached at the nondrive end of the rotor to force external cooling air to cool the rotor and stator coils. Figures 9.1(a) and (b) show the rotor and stator of an electric motor. The frame has to have adequate strength and rigidity to withstand the reaction forces and loads on the motor, without causing any relative displacements between the stator and the rotor coil, because the air gap is the most important source of noise and vibration in an electric motor.

FIGURE 9.1
(a) Rotor of an electric motor with two bearings at the end, (b) Stator of an electric motor.

9.3 Faults in Electric Motor

Faults in an electric motor are due to three important components—the rotor, stator, and the motor bearings. The rotor in an electrical motor is designed to have a uniform air gap all around it with the stator of the motor. When bearings have defects, the shaft is bent, or if during installation the rotor is not positioned at the center, the air gap becomes eccentric. This air gap eccentricity leads to pulsating radial electromagnetic forces, which leads to an increase in the mechanical vibration of the motor as well as an increase in the presence of additional frequency components and their harmonics in the stator current. The air gap eccentricity can be static or dynamic. The static eccentricity is a result of the fact that the position of the minimum air gap is fixed in space. In dynamic eccentricity, the position changes with time. In dynamic eccentricity, the center of the rotor is not at the center of the rotation.

The coils in the rotor are in the form of bars inserted in slots on the rotor core with end rings. These rotor bars can develop cracks and break due to differential thermal expansion and unequal magnetic forces.

The stator coils are electrically insulated; on being subjected to thermal temperature cycling, excessive temperature, and high-frequency electrical sparks, the insulation may degrade and a stator coil short circuit can occur. The bearings in the motor can have defects due to fluting in the races because of high-frequency electrical discharges, or the presence of impurities in the bearing races due to baking of the bearing lubricant at high temperatures.

9.4 Fault Detection in Electric Motors

The faults in electrical motors can be detected by conventional vibration monitoring at the bearing locations. Usually, the vibration levels at twice the supply line frequency increase with the severity of the defect. The vibration levels can be measured to evaluate the electric motor's health condition depending upon the ISO guideline. With vibration, the bearing temperature is also monitored, since if the bearing temperature increases it could lead to baking of the semisolid lubricant, which would then become a hard abrasive particle that could scour the bearing races. Usually, a safe limit for the motor bearing temperature is 70°C.

Another technique of detecting the condition of induction motors is by stator current signature analysis, which is also known as motor current signature analysis (MCSA). This technique is advantageous because one need not mount any transducer on the motor bearing to know its health condition, as is done in the case of contact-type vibration monitoring using accelerometers. For MCSA, the current drawn by the stator of the motor can even be measured at distant locations from the motor as long as there is access to the current-carrying conductor to the motor. Once the current time waveform is captured, it can be analyzed in the frequency domain. Faults like rotor eccentricity, rotor broken bar, and motor bearing fault, can be easily identified by their characteristic frequencies in the frequency domain.

9.5 MCSA for Fault Detection in Electrical Motors

9.5.1 Broken Rotor Bar

When a broken bar is present within the rotor, current cannot flow through it, and therefore, it can no longer add its share of torque to the rotor's load burden. As the broken bar passes under the pole, it will effectively reduce the torque of the rotor for the period of time it is under the field pole, in its torque-producing position.

There are many reasons for rotor bar and end-ring breakage. They can be caused by the following:

1. Thermal stresses due to thermal overload and unbalance, hot spots, or excessive losses, sparking (mainly fabricated rotors)
2. Magnetic stresses caused by electromagnetic forces, unbalanced magnetic pull, electromagnetic noise and vibration
3. Residual stresses due to manufacturing problems
4. Dynamic stresses arising from shaft torques, centrifugal forces, and cyclic stresses
5. Environmental stresses caused by, for example, contamination and abrasion of rotor material due to chemicals or moisture
6. Mechanical stresses due to loose laminations, fatigued parts, bearing failure, and so on

Due to the presence of broken rotor bars, the motor may not start because it may not be able to develop sufficient accelerating torque. Regardless, the presence of broken rotor bars precipitates deterioration in other components, which can result in time-consuming and expensive fixes. Broken rotor bars are detected by monitoring the motor current spectral components produced by the magnetic fields anomaly of the broken bars.

The frequency of the sidebands due to a broken rotor bar is given by Equation (9.1)

$$f_{brb} = f_s \left[k \left(\frac{1-s}{p} \right) \pm s \right] \qquad (9.1)$$

where k/p = 1, 3, 5, 7...., and s is the slip given as $s = \dfrac{f_s - f_r}{f_s}$.

9.5.2 Eccentricity-Related Faults

Machine eccentricity involves an unequal air gap that exists between the stator and rotor. If the air gap between the stator and rotor is not uniform, the forces on the rotor are not balanced, resulting in high magnetically induced vibration at twice the supply frequency. The magnetic attraction is inversely proportional to the square of the distance between the rotor and stator, so a small eccentricity causes a relatively large vibration.

There are two types of air gap eccentricity:

1. Static air gap eccentricity
2. Dynamic air gap eccentricity

Static eccentricity may be caused by any of the following:

1. Ovality of the stator core
2. Out-of-tolerance spigots in the end frames that house the bearings

3. Incorrect positioning of the rotor or the stator at the commissioning stage

4. Incorrect installation of large high voltage (HV) motors. As the motor is mounted to the base, the motor housing and stator are distorted, resulting in the uneven air gap between the stator and the rotor.

Dynamic eccentricity causes the following:

1. Nonconcentric rotor and shaft
2. Thermal bow of the rotor
3. Combination of high air gap eccentricity, consequential unbalanced magnetic pull (UMP), and a flexible rotor
4. Severe bearing wear
5. Rotor core lamination movement independent of the shaft

The dynamic eccentricity may be caused by several factors:

1. Bent rotor shaft,
2. Bearing wear or
3. Misalignment or
4. Mechanical resonance at critical speed, etc.

Generally, both static and dynamic eccentricities coexist in the motor. The presence of the static and dynamic eccentricity can be detected using MCSA. Frequency of the sidebands due to air gap eccentricity is given by

$$fe = f_s \left[\left(kR \pm n_d \right) \frac{(1-s)}{p} \pm v \right] \tag{9.2}$$

where, k = 1, 2, 3 .... And v = ±1, ±3, ±5, and so on.

Flux density distribution varies in both space and time, and the time component is represented by dynamic eccentricity index.

Eccentricity order number = 1 {for dynamic eccentricity}.

For static eccentricity n_d = 0; therefore, Equation (9.2) can be written as:

$$fe = f_s \left[\left(kR \right) \frac{(1-s)}{p} \pm v \right] \tag{9.3}$$

9.5.3 Bearing Faults

When there are faults in the motor bearings, sidebands around the supply frequency indicate the presence of bearing faults. The bearing defect frequency is given in Equation (9.4), where $f_{i,o}$ is the characteristic bearing frequency, which depends on the bearing configuration and dimensions, described in Chapter 7.

$$f_{bng} = |f_s \pm mf_{i,o}| \tag{9.4}$$

9.6 Instrumentation for Motor Current Signature Analysis

MCSA is a nonintrusive fault identification method. The current drawn by the stator is measured and captured by a Hall effect current sensor as shown in Figure 6.13. A current transformer coil can also be used, however they are not suitable for low-frequency measurements. Once the current waveform is captured, it can be analyzed in a frequency analyzer and the frequency characteristics of the stator current obtained as per the schematic diagram shown in Figure 9.2. The instrumentation required for MCSA are of relatively less cost compared to the conventional vibration-based measurements using accelerometers.

A three-phase electric motor driven by a variable frequency drive was operated at a synchronous speed of 20 Hz. The stator current spectrum for the normal motor is shown in Figure 9.3. The motor was then replaced with

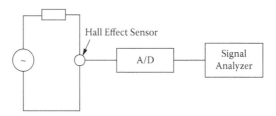

FIGURE 9.2
Schematic diagram for motor current monitoring.

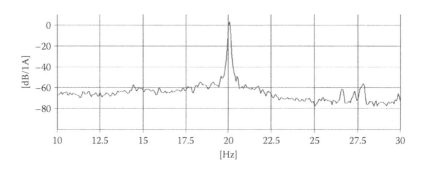

FIGURE 9.3
Stator current spectrum for normal motor at 20-Hz supply frequency.

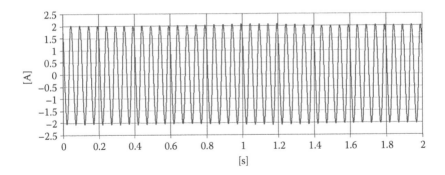

FIGURE 9.4
Time history of stator current for broken rotor bar case.

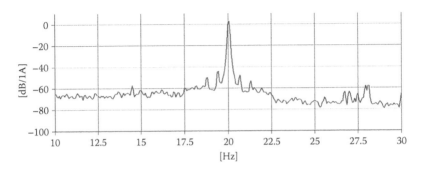

FIGURE 9.5
Stator current spectrum for broken rotor bar in motor.

a defective motor with three broken bars. The modulated stator current due to the broken rotor bar is given in Figure 9.4, where around the time of 1 second, the increase in the current amplitude can be seen due to modulation. The frequency spectrum of the stator current for the defective motor shown in Figure 9.5 shows the presence of sidebands around the supply frequency of 20 Hz. In this case, the rotor speed was measured to be 1184 RPM.

9.7 Fault Detection in Mechanical Systems by MCSA

A significant advantage of MCSA is the fact that the faults in mechanical systems being driven by the electric motors can be indirectly detected by monitoring the stator current. This is because a defect in the mechanical system like a motor operated valve, gear box, pump, and so on, will produce a load torque on the rotor of the motor which will induce a magnetomotive force in the stator coils. The current in the stator coil will thus be modulated and sidebands around the supply frequency spaced at the mechanical defect frequencies will be observed. In order to identify these mechanical frequencies, an amplitude demodulation or envelope analysis of the stator current is done to identify the mechanical defect frequencies.

9.7.1 Relation between Vibration and Motor Current

When an induction motor is driving a mechanical system, the input shaft of the mechanical system is coupled rigidly to the rotor of the induction motor. In such a case, the load torque on the rotor will be a function of the input shaft speed and its modulated frequencies, which will affect the electromagnetic field of the stator. The stator will draw current according to the variation in the electromagnetic field.

The air-gap torque (T) in the induction motor consists of a constant (or average) torque and some oscillatory torques due to torsional vibrations at frequencies f_1, f_2, and f_3 with respective phases of ϕ_1, ϕ_2, and ϕ_3 given by Equation (9.5).

$$T = T_0 + T_1 \cos(2\pi f_1 t + \phi_1) + T_2 \cos(2\pi f_2 t + \phi_2) + T_3 \cos(2\pi f_3 t + \phi_3) \quad (9.5)$$

The current in any phase of the induction motor will have two components, magnetizing current component I_{sM}, which is in phase with flux vector; and a torque producing component I_{sT}, which is 90° ahead of the flux vector. The vector diagram is shown in Figure 9.6. Due to torsional vibration, these

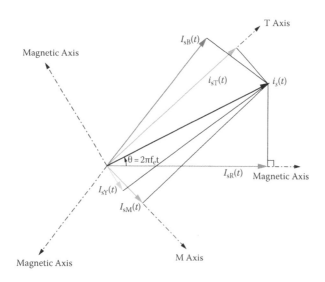

FIGURE 9.6
Vector diagram of all current components in the induction motor.

components will also have an average value such as I_{sM0} and I_{sT0}, and some oscillating components given in Equation (9.6).

$$I_{sM} = I_{sM_0} + A_{sM_1} \sin(2\pi f_1 t + \phi_{M_1}) + A_{sM_2} \sin(2\pi f_2 t + \phi_{M_2})$$
$$+ A_{sM_3} \sin(2\pi f_3 t + \phi_{M_3}) \tag{9.6 a-b}$$
$$I_{sT} = I_{sT_0} + A_{sT_1} \cos(2\pi f_1 t + \phi_{T_1}) + A_{sT_2} \cos(2\pi f_2 t + \phi_{T_2})$$
$$+ A_{sT_3} \cos(2\pi f_3 t + \phi_{T_3})$$

The current drawn by the induction motor would have been a pure sinusoidal wave, had there been no defect in the induction motor and no oscillation in the load torque. But due to the presence of these oscillating frequencies I_{sT} and I_{sM}, the waveform of the current will be affected. The current in the R phase is given by the following equation.

$$I_{sR} = I_{sM} \sin 2\pi f_e t + I_{sT} \cos 2\pi f_e t$$

$$= \left[\begin{array}{l} I_{sM_0} + A_{sM_1} \sin(2\pi f_1 t + \phi_{M_1}) + A_{sM_2} \sin(2\pi f_2 t + \phi_{M_2}) \\ + A_{sM_3} \sin(2\pi f_3 t + \phi_{M_3}) \end{array} \right] \sin 2\pi f_e t$$

$$+ \left[\begin{array}{l} I_{sT_0} + A_{sT_1} \cos(2\pi f_1 t + \theta_{T_1}) + A_{sT_2} \cos(2\pi f_2 t + \theta_{T_2}) \\ + A_{sT_3} \cos(2\pi f_3 t + \theta_{T_3}) \end{array} \right] \cos 2\pi f_e t \tag{9.7}$$

Simplifying the above equation,

$$I_{sR} = I_0 \sin(\sin 2\pi f_e t + \phi_0)$$

$$+ \left(\frac{A_{sT_1} + A_{sM_1}}{2} \right) \cos(2\pi(f_e - f_1)t - \phi_{M_1})$$

$$+ \left(\frac{A_{sT_1} - A_{sM_1}}{2} \right) \cos(2\pi(f_e + f_1)t + \phi_{M_1})$$

$$+ \left(\frac{A_{sT_2} + A_{sM_2}}{2} \right) \cos(2\pi(f_e - f_2)t - \phi_{M_2}) \tag{9.8}$$

$$+ \left(\frac{A_{sT_2} - A_{sM_2}}{2} \right) \cos(2\pi(f_e + f_2)t + \phi_{M_2})$$

$$+ \left(\frac{A_{sT_3} + A_{sM_3}}{2} \right) \cos(2\pi(f_e - f_3)t - \phi_{M_3})$$

$$+ \left(\frac{A_{sT_3} + A_{sM_3}}{2} \right) \cos(2\pi(f_e + f_3)t + \phi_{M_3})$$

where $I_{sM} = I_0 \cos\phi_0$; $I_{sT} = I_0 \sin\phi_0$ and $\phi_0 = \tan^{-1} \dfrac{I_{sT}}{I_{sM}}$. Here it is assumed that $\phi_M \cong \phi_T$.

The stator current components in the Y phase and B phase can be expressed by Equation (9.9 a-b). Putting Equation (9.6 a-b) in Equation (9.9 a-b), expressions similar to Equation (9.8) can be derived.

$$I_{sY} = I_{sM} \sin\left(2\pi f_e t - \frac{2\pi}{3}\right) + I_{sT} \cos\left(2\pi f_e t - \frac{2\pi}{3}\right) \quad \text{and}$$

$$I_{sB} = I_{sM} \sin\left(2\pi f_e t + \frac{2\pi}{3}\right) + I_{sT} \cos\left(2\pi f_e t + \frac{2\pi}{3}\right)$$

(9.9 a-b)

The result indicates that the oscillating load torque with specific frequencies will induce sidebands of these frequencies across the supply line frequency in the current signature drawn by an induction motor. This load torque will subsequently give rise to fluctuation in speed, which is simply the torsional vibration of the shaft. In the mechanical system connected to an induction motor, the torsional vibration response will indicate the sidebands across the supply line frequency in the current signatures.

9.7.2 Fault Detection in a Submersible Pump

A submersible pump with seeded faults in the impeller was used for fault detection using MCSA. The normal impeller and an impeller with one broken vane of the submersible pump are shown in Figure 9.7. The schematic of the experimental setup is shown in Figure 9.8. An underwater accelerometer was also used to measure the vibration of the pump to correlate the fault detection by MCSA, as shown in Figure 9.8.

The stator current spectra for the normal and defective submersible pump are shown in Figures 9.9 and 9.10. In the spectrum of the defective impeller in the submersible pump, sidebands around the harmonics of the supply frequency are prominent.

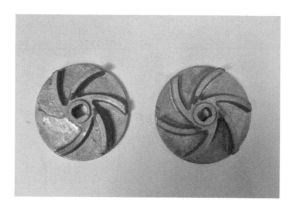

FIGURE 9.7
Broken and normal impeller of submersible pump.

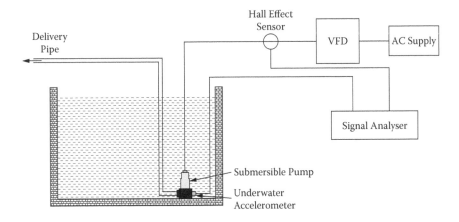

FIGURE 9.8
Schematic line diagram of the experimental setup for MCSA.

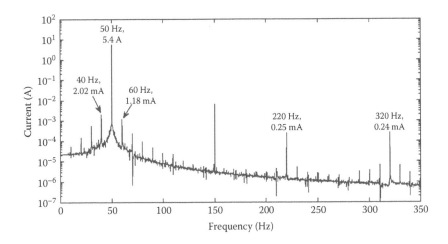

FIGURE 9.9
Stator current spectrum for normal pump.

9.8 MCSA for Fault Detection in Any Rotating Machine

MCSA is a powerful technique for determining faults in mechanical systems that are otherwise inaccessible for health monitoring by conventional vibration analysis. For example, a submersible pump that is buried underground is not accessible for mounting an accelerometer, or pumps and valves inside a nuclear reactor that are driven by an electric motor are excellent candidates for employing MCSA for fault detection. In MCSA, the electric

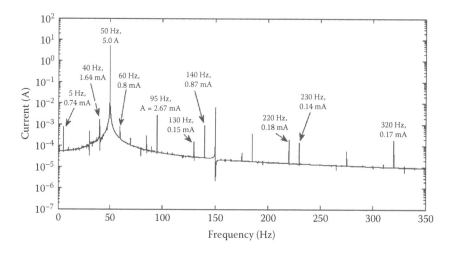

FIGURE 9.10
Stator current spectrum for defective pump.

motor itself behaves as a transducer. Going a step further, if any rotating mechanical system is attached to a tacho-generator, the frequency of ripples in the voltage generated by a tacho-generator is at the frequency of the mechanical component and any mechanical defects associated with it. Many of the gas turbine– based propulsion systems usually have a generator coupled to them to produce electrical energy for driving onboard propulsion motors. MCSA can be used in such systems for online fault detection from a remote location, even without any personnel going close to the mechanical system. This quality of MCSA has made it an ideal choice for health monitoring of windmill gearboxes as well.

9.9 Fault Detection in Power Supply Transformers

In electrical power distribution systems, step-down transformers are a very important component. The essential elements of a step-down power transformer are a primary and a secondary coil wound around soft iron laminates, insulated and placed in a tank with transformer oil around it. The primary purpose of the transformer oil is to cool the coils from the eddy current heat generated due to the magento-restrictive effect in the transformer due to electromagnetic induction. The coils are laminated to reduce the eddy current losses in the transformer. The tank of the transformer is made in such a way that there are fins around it for the oil to pass through and exchange the generated heat by natural convection to the surroundings. Usually, transformer failures occur when the heat generated cannot be efficiently liberated, which results in the breaking of the insulation layer due to thermal effect and/or loss of the dielectric strength of the transformer oil. A sight glass is usually provided in the transformer oil tank to check the level of the oil in the tank, and to ensure that the required minimum quantity of the transformer oil is present. Electrical utility companies routinely do a transformer oil analysis to check for any deterioration of the oil's properties.

9.10 Fault Detection in Switchgear Devices

In electrical power distribution switchyards, power distribution switchboards, and electrical control panels, the occurrence of loose electrical connections is responsible for loss in electrical power supply. Many electrical utility companies have minimized this power loss by ensuring that loose connections between two mating current-carrying conductors do not occur. The loose connection essential provides a gap between two electrodes, and depending on the potential difference across them, an electrical spark is generated. This repeated sparking causes a heat buildup, and if it goes undetected, can cause a fire if flammable objects are around the sparking conductors. To reroute the electrical power flow in electrical power transmission lines, manual switches are operated through a level from the base of the tower to make contact with a set of conductors to direct the power flow. However, again, if a loose contact occurs, sparking would occur resulting in heat generation and power loss. Thus, in order to detect heat generation in an electrical switch gear, temperature measurements using infrared thermography are done. Details about thermography are presented in the next chapter.

10

Thermography

10.1 Introduction

Thermography is a thermal imaging technique using an infrared camera.

Any material that is at a temperature above 0 K (–273.16°C) gives out electromagnetic radiation at a particular wavelength. The wavelength of this electromagnetic radiation is just above the wavelengths of visible light. The wavelength of the visible light is 4000 to 7000 angstrom (1 angstrom = 10^{-10} m). Electromagnetic waves radiated in the range of 9000 to 14,000 angstrom are known as (IR) infrared waves. Figure 10.1 gives a relative distribution of the electromagnetic waves based on their wavelength of radiation. This thermal radiation is measured by suitable infrared detectors to provide a means to know the surface temperature of the thermal energy-emitting body. However, there needs to be a temperature difference between the source and the location where the radiated thermal energy is being captured, so that the thermal energy can flow. The energy emitted by the body is directly proportional to the fourth power of its temperature in Kelvin scale. This constant of proportionality is known as the Stefan–Boltzmann constant, which has a value of 5.67×10^{-8} W/m^{2}K^4.

The total radiated thermal energy, W can be described by Equation (10.1).

$$W = \alpha W + \rho W + \tau W \qquad (10.1)$$

The fractions of the radiated energy, which describe the absorption of the incident energy, is denoted by α, the reflected energy is ρ, and the transmitted energy is τ. Each coefficient can vary anywhere from zero to one. A body where no energy is transmitted or reflected and all the incident energy is absorbed is known as a *black body*. This amount of energy absorbed by a black body is independent of the wavelength of thermal radiation. A black body is also a perfect emitter and the energy emitted by a black body is independent of the direction of radiation. However, no real-world body is a perfect black body and thus every body that radiates thermal energy is characterized by its surface emissivity, given by Equation (10.2).

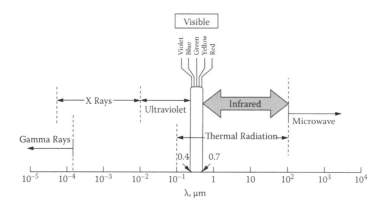

FIGURE 10.1
Electromagnetic spectrum.

$$\varepsilon = \frac{W_{obj}}{W_{bb}} \tag{10.2}$$

where W_{obj} and W_{bb} are the heat emitted by the object and a black body, respectively. The emissivity of some commonly used materials are given in Table 10.1.

Usually, shiny metals have low emissivity values. An object that has the same emissivity at all wavelengths is known as a grey body. Thus the Stefan-Boltzmann law is given as in Equation (10.3)

$$W = \varepsilon \sigma T^4 \tag{10.3}$$

Based on the above principles of thermal radiation, commercial infrared detectors are available for measuring the surface temperature of a radiating body.

10.2 Thermal Imaging Devices

The thermal energy emitted by a black body varies according to the wavelength of the radiation and is given, according to the Plank distribution, by Equation (10.4).

$$E_{b\lambda} = \frac{C_1 \lambda^{-5}}{e^{C_2/\lambda T} - 1} \tag{10.4}$$

where
   $E_{b\lambda}$ = monochromatic black body emissive power [W/m²·µm]
   λ = wavelength, µm
   T = temperature, K
   $C_1 = 3.743 \times 10^8$ W·µm⁴/m²
   $C_2 = 1.4387 \times 10^4$ µm·K

TABLE 10.1

Typical Emissivity Values of Some Common Materials

Surface Material	Emissivity Coefficient
Aluminum heavily oxidized	0.2–0.31
Aluminum paint	0.27–0.67
Black body matt	**1.0**
Brass polished	0.03
Brick, red rough	0.93
Concrete	0.85
Cotton clothes	0.77
Copper electroplated	0.03
Glass Pyrex	0.85–0.95
Granite	0.45
Gypsum	0.85
Iron polished	0.14–0.38
Limestone	0.9–0.93
Mild steel	0.20–0.32
Plaster	0.98
Paper	0.93
Plastic	0.91
Rubber hard glossy plate	0.94
Sand	0.76
Stainless steel, weathered	0.85
Tile	0.97
Water	0.95–0.963
Wood, pine	0.95
Wrought iron	0.94

From Equation (10.4), it is obvious that the radiant energy is dependent on the surface temperature.

10.2.1 Optical Pyrometer

Optical pyrometers are used to measure surface temperature of radiating bodies, where, because of the high temperatures (as in the blast furnace of a steel plant), conventional contact-type measurements at high temperatures above 500°C would not work. These pyrometers are designed so that the radiated thermal waves are focused on a screen. Another secondary circuit is present inside the pyrometer that is responsible for heating the filament inside it and is passed through an optical red filter. By varying the current to the filament, the intensity of the radiated heat can be varied. By superimposing the radiant wave from the body whose temperature is to be measured, to the image formed on the screen, the temperature can be estimated once the image of the filament vanishes from the image screen.

These optical pyrometers are handheld devices and require manual focusing. The usual range of these pyrometers are from 850°C to 1200°C with an accuracy of ±5°C.

10.2.2 Infrared Cameras

The infrared cameras used for thermography are similar to digital video cameras. However, a major difference of these IR cameras over a home digital video camera is that the IR camera uses focal plane arrays (FPAs) to detect micrometer-sized pixels of materials that are sensitive to IR wavelengths instead of the charge-coupled device (CCD) in a home video camera. The main components of an IR camera are a lens that focuses IR onto a detector, associated electronics, and software for processing and displaying signals and images. The resolution of the FPA can be 1024 × 1024 pixels. IR cameras have built-in software that allows the user to focus on a particular area of the FPA and calculate the temperature. IR cameras have temperature measurement precision better than ±1°C. IR cameras typically use silicon and germanium materials in the FPA.

A typical hand-held IR camera used for quick surveys has a 120 × 240 pixel resolution. These IR cameras have onboard memory to save the thermal images of the object whose temperature is measured. The operator of such an IR camera usually enters the surface emissivity of the surface whose temperature is to be measured.

10.3 Use of IR Camera

In order to use an IR camera, the following should be noted. The camera has to be focused on the object whose temperature is to be measured. In many cameras, an optional laser is provided to aid in targeting and focusing the object. Care must be taken that no thermal emissions from any unwanted glare or reflections enter the camera. Most important of all is to set the correct emissivity value in the camera so that the software can calculate the correct temperature based on the amount of thermal radiation. An in situ calibration technique for the camera is suggested. The camera is usually focused on another similar color surface whose temperature is known; then an emissivity value is entered in the software until the correct temperature is displayed, though many times IR camera users obtain a value of the surface emissivity from handbooks or other sources; one such set of values is provided in Table 10.1.

The IR camera has built-in capability to transfer data via USB to a computer. The rate of data transfer is either 30 image frames per second or 25 frames per second, though the image transfer rate can be controlled by software.

High image transfer rates help in capturing several transient heat-transfer phenomena. The image captured from an IR camera can be analyzed by commercially available software to determine the distribution of temperature in a particular area of the image, and necessary statistical features of such images can also be estimated.

10.4 Industrial Applications of Thermography

Thermography has found many industrial applications for two obvious reasons: First, thermography is noncontact in nature, and temperature can be measured from a distant location. Second, light, portable, and precise IR image cameras are commercially available. The recent advances in instrumentation along with the availability of user-friendly software has made thermography useful in many engineering applications. Figure 10.2 shows a hand-held thermal imaging camera being used for temperature monitoring of an engine.

10.4.1 Leakage Detection

In many types of industrial process equipment, hot fluids and gases flow in pipelines. Thermal imaging of such pipelines is done to detect leakage at seals and joints. Proper corrective measures can thus be taken to stop such unwarranted leakage.

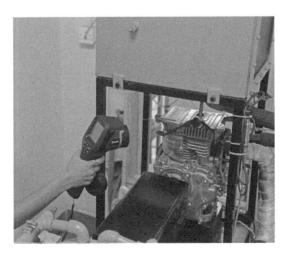

FIGURE 10.2
Handheld thermal imaging camera being used for engine temperature monitoring.

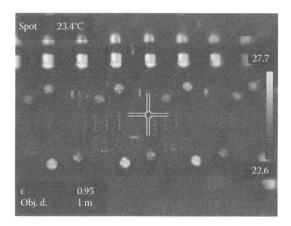

FIGURE 10.3
IR image of an electronic amplifier.

10.4.2 Electrical and Electronic Component Heat Generation

In many electrical and electronic systems and circuits, due to flow of current in a conductor, heat is generated that is proportional to the square of the current. In many instances, due to an increase in the load resistance, the current in the circuit increases as a result of which there is a temperature rise, which can be easily detected with infrared imaging. The increase in resistance could be due to oxide formation at the electrical contact points, electrical sparking due to loose contacts, and so on, which in turn increases the current flowing through the conductors. In a plant control room or a system of machineries where several electrical conductors connect the various power sources with electrical loads through switchgears, thermography aids in a quick survey to detect a conductor at a higher temperature. Electrical motor defects like broken rotor bars, stator short circuits, and so on, can also be detected by thermal imaging. Figure 10.3 shows the heated electronic components in an amplifier detected by thermal imaging.

10.4.3 Building Condition

In many instances, several defects can occur in buildings, like roof leakage, cracks at building joints, mold growth due to dampness, and so on. These result in different temperatures in different parts of the building. Through infrared imaging of such building structures, variations in temperature can be measured and clues as to cause of the building defect can be obtained.

10.4.4 Machineries

IR imaging is used for thermal imaging of processing equipment. Figure 10.4 shows the thermal image of a reciprocating compressor. It can be seen that

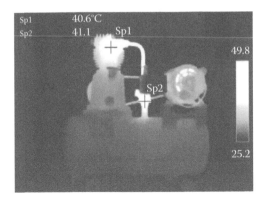

FIGURE 10.4
IR image of a reciprocating compressor.

the compressor cylinder walls are at a much higher temperature than the compressed air reservoir at the bottom.

10.5 Applications of Thermography in Condition Monitoring

Shafts are important machine components that are supported on bearings. The bearings are designed to provide low frictional resistance to the rotating shafts and to support the loads on the shafts. Due to defects in the rotating machines, many times the loads on the shaft increase, which in turn increases the normal reaction at the bearings. Due to an increase in the normal load, the components in the bearing undergo deflection along with an increase in frictional torques. The heat generated at the bearings thus increases, which is responsible for the increase in temperature at the bearings. The rate of heat generation is proportional to the rotational speed of the shafts. With an increase in the generated heat at the bearings, the grease that is present to lubricate the bearings liquefies and flows out. Due to excessive heat generated at the bearings, the remainder of the grease can become baked and form hard particles. These particles score the bearing races, which leads to further deterioration of the bearing. To extend the life of the bearings, the temperatures of rolling element bearings are not allowed to rise above 70°C.

Infrared thermal imaging can be used to monitor the temperature of bearings, particularly in distant locations. Similar to the phenomenon of heat rise in rolling element bearings, flexible couplings, in which flexure elements are present in the coupling, experience frictional heat generation because of excessive forces at the couplings due to shaft misalignment. This heat generation is responsible for the increase in the coupling temperature.

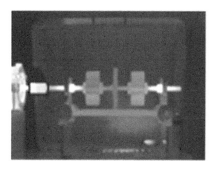

FIGURE 10.5
Thermal image of the shaft coupling.

When the shaft is in motion, traditional temperature measurements at the shaft couplings using thermocouples is not convenient. In such situations, the temperature of the coupling can be measured through infrared thermal imaging. Figure 10.5 shows the thermal image of the temperature of the coupling while in motion at the end of 120 seconds from the start of the machine system with misaligned shafts. It has been found by infrared thermal imaging that the rate of temperature rise of couplings in the case of misaligned shafts is more than that in the normal case with no shaft misalignment.

11

Wear Debris Analysis

11.1 Introduction

Machines have many moving parts that rub against each other during operation. These moving parts are subjected to dynamic forces, and due to friction between the parts, energy is dissipated as heat. Machine designers and manufacturers take care to ensure that these frictional forces are minimized, so that energy is not unnecessarily dissipated. Surfaces that rub or mate against each other are heat treated so that the surface is hard and wear resistant, and a layer of lubricant is applied to reduce friction and dissipate whatever frictional heat is generated. This aspect of the machine is considered in its condition monitoring. No matter how wear resistant the surfaces are, with time, particles from the surface of the mating machine members will be dislodged by a wear phenomenon and will be deposited in the lubricating oil. The concentration of such wear particles, their size, shape, and composition provide clues to the maintenance engineer regarding the health of the machine. Further, due to the deposition of wear particles in the lubricant (which may be oil or grease), the physical and chemical properties of the lubricant change. Thus, by measuring these properties of the lubricating oil, an indication of the machine's health can be obtained. This chapter presents an overview of the mechanisms of wear and the analysis of wear particles and lubricating oil.

11.2 Mechanisms of Wear

When machine components rub against each other, there may be relative sliding or rolling motion between the parts. These forces and the surface condition of these parts are responsible for creating wear particles. Wear that occurs between mating machine components is classified into four categories:

  i. Adhesion wear
  ii. Abrasive wear

iii. Corrosive wear

iv. Fatigue wear

11.2.1 Adhesive Wear

This type of wear occurs when two machine surfaces rub against each other, with or without any lubricant between them. Adhesive wear occurs because no matter how smooth the surfaces are, there will be asperities on the surface at the submicron level. This is usually quantified by the surface roughness value. Under compressive loads at the two mating surfaces, these asperities lock into each other and may become welded due to high pressures and be sheared off during the sliding motion. Thus, material is scoured out at the surface, which may leave behind larger pits. This type of wear is thus also known as *scouring, gaffing,* or *galling* wear.

11.2.2 Abrasive Wear

This type of wear occurs when two materials of different hardness rub against each other. The harder material under load ploughs or cuts into the softer material and pulls out wear particles. Abrasion also occurs if a third hard material (dirt or impurity) is trapped in between two rubbing surfaces. This type of wear is also known as *cutting wear*. A typical example of abrasive wear is a journal bearing with a Babbitt lining, which is softer than the hard steel shaft that is supported by it. The hard impurity between the two surfaces could be due to hard dirt particles suspended in the air, for example, the coal dust in a material-handling system at a thermal power plant. Figure 11.1 shows a third foreign particle trapped between two rubbing surfaces.

11.2.3 Corrosion Wear

Corrosive wear occurs due to the presence of hard particles, usually oxides or hydrates, between two surfaces that slide against each other. These particles are responsible for the surface abrasion. A layer of lubricant on the

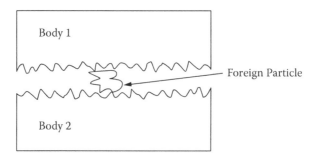

FIGURE 11.1
Trapped foreign particle.

surface protects the material from exposure to air or moisture, so that oxides or hydrates are not formed. However, if the proper amount and type of lubricant is not used, there may be oxide formation, which can be responsible for corrosion wear. Further, when two dissimilar metals rub against each other, electrochemical reaction may also lead to the formation of salts that can rub against the surfaces during the sliding motion and cause surface abrasion.

11.2.4 Fatigue Wear

When two surfaces rotate against each other, the loads on them also change, creating a fatigue load at the contact points of the two mating surfaces. The stress at the contact points can be estimated by Hertzian stress theory. Due to fatigue load, the stress that can be withstood by the material greatly reduces, and the shear stress at small depths below the surface is less. Thus, during rotation, particles suddenly break out from under the surface and create pits. Fatigue wear may be sudden and catastrophic, unlike the other three wear types which are gradual and their wear rate is proportional with time.

11.3 Detection of Wear Particles

Wear particles or debris generated during sliding or rolling contact between machine parts usually get lodged in the lubricating oil. Wear particles are of different sizes, shapes, and chemical composition. Depending on the size of the wear debris in the oil, different detection techniques are used. The common classification of wear debris particles is based on the particle size and means of detection:

   i. Spectroscopy
  ii. Ferrography
 iii. Particle count

Wear debris analysis cannot be done in situ, unlike vibration signature analysis. Analysis of wear debris that is suspended in lubricating oil is done in a dedicated laboratory. Samples of the oil and suspended particles are collected based on specific procedures and sent to a laboratory. Depending on the size of the debris, sample-appropriate analysis techniques are used.

11.3.1 Spectroscopy

The size of wear debris particles that are deposited in lubricating oil ranges from submicrons to a few hundred microns. These particles are suspended in the lubrication oil and are so small that they cannot be clearly seen by the naked eye. For spectroscopy analysis, the particle size is less than 2 microns. To estimate the level of wear contaminants in a given volume of oil, spectroscopy analysis is done to determine the concentration of a particular chemical element in parts per million (ppm) in the sample. There are several spectral analysis techniques that are commonly used such as atomic absorption spectroscopy, atomic emission spectroscopy, X-ray fluorescence, and so on. It is helpful to know the chemical composition of machine components as manufactured. The machine component that has a higher wear rate can then be identified by periodically monitoring the concentration of a particular chemical element in the contaminated oil sample. For example, increase in the concentration of Pb in a Babbitt-lined journal bearing indicates a high rate of wear in the journal bearing.

Optical microscopy is also used to identify the nature of wear in the machine, by monitoring the shape of the debris particles. A scanning electron microscope (SEM) is usually used to view such submicron-sized debris. A scanning electron microscope is shown in steel Figure 11.2. The SEM images of rolled steel before and after wear is shown in Figures 11.3a and b.

11.3.2 Ferrography

For wear debris in the size range of 10 to 100 microns, the particles are too big for spectroscopic analysis. An analysis based on magnetic field is used

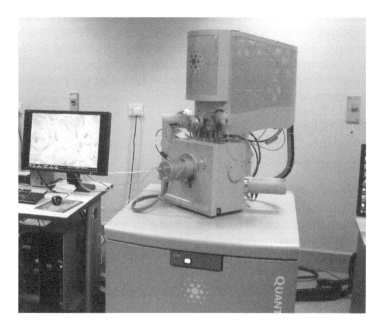

FIGURE 11.2
Scanning electron microscope.

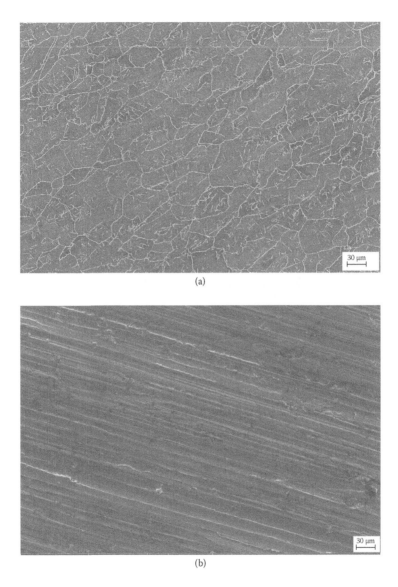

FIGURE 11.3
SEM image of rolled steel before (a) and after (b) wear.

to separate the ferrous particles in an oil sample. There are two types of ferrographic analysis—direct ferrography and analytical ferrography.

In direct ferrography, a contaminated oil sample is siphoned through a glass tube. Debris particles are positioned at different locations along the length of the tube, depending on the size of the particles. By using optical sensors, the amount and distribution of each particle size can be obtained. In analytical ferrography, a technique based on the gravity field is used. A contaminated oil sample is poured down an inclined glass plate and subjected to a magnetic field from behind. Ferrous particles lie on the glass plate with the heavier particles at the bottom of the inclined plate followed by the lighter debris particles. These substrates are used to prepare ferrograms, which can be analyzed using an optical microscope. Ferrography has traditionally been used in the offline mode.

11.3.3 Particle Count

When the size of the wear debris particles in the oil becomes greater than 100 microns, it is easier to count them in a given volume of the contaminated oil with a particle count meter. A light beam is focused on the suspension of the wear debris in an oil sample, and a photo sensor is used to detect the intensity of the light that is transmitted through the suspension. The intensity of the light beam is related to the amount of wear debris present in the oil. The light is made to pass through filters of different sizes, thus a distribution of the wear debris particle size against the amount of debris present is obtained.

In some instances, when sizes of the ferrous debris particles are large, magnetic chip detectors are also used.

11.4 Common Wear Materials

In many engineering applications, machinery components are made up of different chemical elements. Prior knowledge of the chemical composition of the engineering component helps one identify the source of the debris. Table 11.1 provides the chemical elements that are present in a few engineering components.

11.5 Oil Sampling Technique

The wear debris that contaminates the lubricating oil is usually analyzed by collecting the oil from a few particular locations in the lubricating system of the machine. Since the wear debris can be of submicron sizes, care has to be taken regarding the cleanliness of the oil collecting and storage system. The locations from which the oil is collected must also represent the true dynamics of the wear mechanism. Therefore, oil must be collected when the engine is running because, due to gravity, all the debris would be deposited in the bottom of the lubricating sump if samples are taken when the machine is idle. Figure 11.4 shows the possible oil collection points in a lubricating

TABLE 11.1

Sources of Chemical Elements Found in Wear Debris

Materials	Likely Machine Component Source
Aluminium (Al)	Light alloy pistons, crankshaft bearing, component rubbing on casings
Antimony (Sb)	White metal plain bearings
Boron (Bo)	Coolant leaks, can be present as an oil additive
Chromium (Cr)	Piston rings or cylinder liners, valve seats
Cobalt (Co)	Valve seats, hard coatings
Copper (Cu)	Copper-lead or bronze bearings, rolling element bearing cages
Indium (In)	Crankshaft bearings
Iron (Fe)	Gears, shafts, cast iron cylinder bores
Lead (Pb)	Plain bearings
Magnesium (Mg)	Wear of plastic components with talc fillers, seawater intrusion
Nickel (Ni)	Valve seat, alloy steels
Potassium (k)	Coolant leaks
Silicon (Si)	Mineral dust intrusion
Silver (Ag)	Silver-plated bearing surfaces, fretting of silver soldered joints
Sodium (Na)	Coolant leakage, seawater intrusion
Tine (Sn)	Plain bearings
Vanadium (V)	Intrusion of heavy fuel oil
Zinc (zn)	A common oil additive

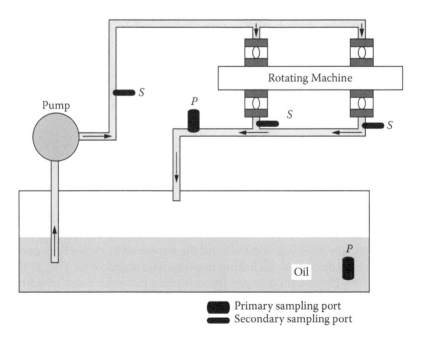

FIGURE 11.4
Preferred oil sampling locations.

TABLE 11.2

Preferred Frequency of Oil Sampling

Machinery	Sampling Interval (hours)
Diesel engine—off highway	150
Transmission, differentials, final drives	300
Hydraulics—mobile equipment	200
Gas turbines	500
Steam turbines	500
Air/gas compressors	500
Chillers	500
Gear boxes—high speed/duty	300
Gear boxes—low speed/duty	1000
Bearings—journal and rolling element	500
Aviation reciprocating engines	25–50
Aviation gas turbines	100
Aviation gear boxes	100–200
Aviation hydraulics	100–200

system for wear debris and oil analysis. A hand-drawn vacuum pump is usually used for siphoning the oil into the sampling bottle.

The cleanliness of the sampling bottle and tubing used to collect oil samples must be maintained per the ISO standards. It is also desirable to use a fresh clean tube at each oil sampling point and to store each sample in a clean bottle. Since the oil is sent to a laboratory for subsequent analysis, proper labeling of the bottle is essential.

In industries where condition monitoring based on wear debris analysis is done, oil sampling is done at a particular frequency. Table 11.2 provides the frequency of oil sampling for different types of machines.

11.6 Oil Analysis

The wear debris from mating machine components is deposited in the lubricating oil. The physical and chemical properties of the lubricating oil also change due to the presence of debris and the temperature cycles they undergo due to machine operation, including intermediate shutdowns. The oil that is collected from the machine as per the sampling procedure described in the earlier section, is tested for its physical and chemical properties. The tests on these sampled oils are done in specialized laboratories as per specified standards. A list of the standards are provided in Appendix A6.

Table 11.3 provides descriptions of some of the properties of the oil that need to be determined in oil analysis.

TABLE 11.3

Oil Properties

Oil Property	Description of the Property
Acidity/alkalinity	Acidity is conventionally measured in terms of the amount of potassium hydroxide (KOH) needed to neutralize the acidity. Alkalinity is the amount of KOH that is equivalent to the acid needed to neutralize alkaline constituents in a fixed quantity (1 gram) of sample.
Corrosion inhibitors	Moisture present in oil or condensing from the atmosphere can lead to corrosion in engines and circulating systems. Rust inhibitors in the oil afford protection through the formation of a thin protective film on metals. Water content test is done to determine the amount of moisture present.
Detergencing	Detergent properties are required to keep combustion and oil-degradation products in suspension and prevent their depositing and baking in high-temperature zones.
Demulsibility	The ability of the oil to separate readily from water which gets to into the lubricant.
Extreme pressure and antiwear	To prevent welding of metal surfaces at point of contact and hence reduce friction.
Flash point	The temperature at which it gives off a flammable vapor in specified conditions.
Oxidation stability	Hydrocarbon oils react with oxygen (in air) to form acid or sludge products. The time taken for oil to oxidize is a function of operating condition, but is highly temperature dependent.
Pour point	An indicator of the ability of an oil or distillate fuel to flow freely at low temperature.
pH value	A measure of the hydrogen ion concentration; indicates whether the fluid/oil is acidic, neutral, or basic in nature.
Specific gravity	The ratio of density of the oil with respect to water. This gives a clue regarding the contamination of the oil.
Viscosity	A measure of a lubricant's resistance to flow.
Viscosity Index (VI)	A measure of the relationship of viscosity to temperature. A low VI means a large variation of viscosity with temperature.

11.7 Limits of Oil Analysis

Oil analysis is routinely done in a specialized laboratory. Table 11.4 provides the ISO code chart for number of particles present in a given milliliter of a fluid/oil. The oil itself can be contaminated with particles of various sizes; ISO standards provide oil cleanliness requirements for different machines, depending upon the criticality of the machine component. The ISO 4406 standard designates a cleanliness level of oil based on the number of contaminant particles of three sizes: 4 µm, 6 µm, and 14 µm, that are present in a milliliter of the oil or fluid. The cleanliness level of the oil as per the ISO standard is expressed in three numbers that correspond to the above three particle sizes.

TABLE 11.4

ISO Code Chart for Number of Particles in a Milliliter of a Fluid

Range	Particle Count More Than	Particle Count More Than
24	80,000	160,000
23	40,000	80,000
22	20,000	40,000
21	10,000	20,000
19	2500	5000
18	1300	2500
17	640	1300
16	320	640
15	160	320
14	80	160
13	40	80
12	20	40
11	10	20
10	5	10

TABLE 11.5

Recommended Cleanliness Level of Oil

Machine Component	ISO Cleanliness Level
Roller Bearings	16/14/12
Journal Bearings	17/15/12
Industrial Gearboxes	17/15/12
Mobile Gearboxes	17/16/13
Diesel Engines	17/16/13
Steam Turbine	18/15/12
Paper Machine	19/16/13
Servo-Valve	13/12/10
Proportional Valve	14/13/11
Variable Volume Pump	15/14/12
Fixed Piston Pump	16/15/12
Vane Pump	16/15/12
Gear Pump	16/15/12
Ball Bearing	14/13/11

With each increase in the ISO range number, the contaminant level doubles. Some of the recommended oil cleanliness levels for machinery component in terms of the contaminant size as per the ISO 4406 standard are provided in Table 11.5. The maintenance personnel have to ensure that the oil cleanliness level is better than the limit indicated in Table 11.5.

12

Other Methods in Condition Monitoring

12.1 Introduction

A few other techniques of machinery condition monitoring are followed in industry (Table 12.1). These techniques are generally known as nondestructive test (NDT) techniques. Out of the entire condition monitoring activities carried out in the world, 10% belong to the NDT category. These techniques are usually done for periodic inspections, failure analysis, quality certification, and regulatory compliance. Some of these techniques require skilled and trained personnel to conduct the tests. Many of these tests are conducted in situ with portable NDT units. A few of the commonly used NDT techniques are visual inspection by specialized optical instruments, dye-penetrant test, magnetic particle testing, eddy current testing, radiography, ultrasonics, and acoustic emission.

12.2 Eddy Current Testing

When an magnetic flux is generated through an electrical coil and the coil is brought to a electrically conductive surface, eddy currents are generated on the surface of the conductive surface. The strength of the eddy currents thus generated depends on the frequency of the electrical voltage in the coil and the magnetic permeability. If there is a defect or surface discontinuity, the eddy currents thus generated are disturbed, which can be sensed through appropriate electrical voltage measurements. This principle is the basis of the use of eddy current probes for surface defect monitoring.

TABLE 12.1

Some NDT Test and Their Capabilities in Detecting Defects

NDT Method	Capabilities
Radiography	• Measures differences in radiation absorption • Inclusions, porosity, cracks
Ultrasonic	• Uses high-frequency sonar to find surface and subsurface defects • Inclusions, porosity, thickness of materials, position of defects
Dye penetrate	• Uses a die to penetrate open defects • Surface cracks and porosity
Magnetic particle	• Uses a magnetic field and iron powder to locate surface and near-surface defects • Surface cracks and defects
Eddy current	• Based on magnetic induction • Measures conductivity, magnetic permeability, physical dimensions, cracks, porosity, and inclusions

A few essential requirements for the generation of eddy current are a high-frequency AC voltage source and a electrically conductive surface. The penetration depth, δ, of the eddy current is given by Equation (12.1),

$$\delta = \frac{1}{\sqrt{\pi f \mu \sigma}} \tag{12.1}$$

where
   f is the test frequency in Hz
   μ is the magnetic permeability in H/mm
   σ is the electrical conductivity in % IACS (International Annealed Copper Standard), where 100% IACS corresponds to a resistivity of 17.20 mΩ/meter for a wire of 1 mm^2 cross section.

Eddy current density does not remain constant across the depth of a material. The density is greatest at the surface and decreases exponentially with depth for a material that is both thick and uniform. The standard depth of penetration is the depth at which the eddy current density is 37% of the material surface value. To detect very shallow depth in a material, and to measure the thickness of the thin sheet, very high frequencies are used. Similarly, in order to detect subsurface defects, and to test highly conductive, magnetic, or thick materials, lower frequencies must be used. Surface eddy current probes are made with coils designed to be driven at relatively high frequencies anywhere from 50 kHz to 500 kHz. The coils in subsurface eddy current probes are designed to be driven at relatively low frequencies of typically 1 kHz to 20 kHz.

Eddy current technology is used for monitoring surface cracks and corrosions, measure the thickness of paint on a metallic surface, and so on.

In order to develop an eddy current scanning probe, several coils can be placed next to each other in a single unit. Eddy current probes are also used to detect surface defects in boiler tubes and heat exchangers.

12.3 Ultrasonic Testing

Ultrasonic testing uses high-frequency sound energy to detect defects. Ultrasonic waves in the frequency range from 100 kHz to 50 MHz are used for nondestructive testing and thickness gauging. Ultrasonic waves can transmit through any material, though the transmission characteristics depend on the acoustical impedance of such materials. These ultrasonic waves can propagate through any material as a longitudinal wave, shear wave, surface (Rayleigh) wave, or Plate (Lamb) wave.

A longitudinal wave is a compression wave in which the particle moves in the same direction as the propagation of the wave.

A shear wave is a wave motion in which the particle motion is perpendicular to the direction of the propagation.

Shear waves have an elliptical particle motion and travel across the surface of a material. The velocity is approximately 90% of the shear wave velocity of the material type and their depth of penetration is approximately equal to one wavelength.

Plate waves have a complex vibration occurring in materials where thickness is less than the wavelength of ultrasound introduced in it.

Various types of ultrasonic probes are available for such testing. A dual-type transducer uses separate transmitting and receiving elements to create a pseudofocus, which is advantageous for inspecting parts with rough backwall surfaces. Applications of such probes include remaining wall thickness measurements, corrosion monitoring, and high-temperature applications.

A contact type ultrasonic probe is indented for direct contact with a test piece. This transducer typically has a hard wear surface optimized for contact with most metals. Applications of such contact-type probes include straight beam flaw detection, thickness gauging, and velocity measurements.

In an angle beam ultrasonic transducer, there is a removable or integral wedge that introduces sound at an angle into the part. The main application of such an angle beam probe is for weld inspection and other flaw detection and crack sizing techniques. Figure 12.1 shows the three most commonly used ultrasonic probes and the types of waves generated in a test material.

A typical ultrasonic testing system consists of a pulser–receiver transducer and a display device. A pulser–receiver is usually a piezoelectric element that generates high-frequency sound pulses in the ultrasonic frequency band. The sound energy is introduced and propagates through the material in the form of waves. When there is a discontinuity (due to a defect), part of the energy is reflected back from the defective surface. The reflected signal is then transformed into an electrical signal by the transducer and is displayed on a screen.

Signal travel time can be directly related to the distance that the signal traveled by knowing the speed of sound in the test material. From the reflected signal information about type size, the orientation and location of the defect can be known.

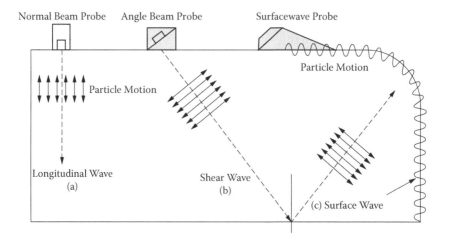

FIGURE 12.1
Common ultrasonic probes.

Ultrasonic is sensitive to both surface and subsurface discontinuities. The depth of penetration for flaw detection or measurement is superior to other NDT methods. In ultrasonic testing, only single-sided access is needed. Minimal part preparation is required. However, it is difficult to perform ultrasonic testing on materials that are rough, irregular in shape, very small, exceptionally thin, or not homogeneous. Cast iron and other coarse-grained materials are difficult to inspect due to low sound transmission and high signal noise. Linear defects oriented parallel to the sound beam may go undetected.

Ultrasonic data is usually presented in three common ways. They are known as the A scan, B scan, and C scan. The A scan presents a two-dimensional plot of the reflected ultrasonic energy as a function of time. The B scan represents a cross-sectional view of the test specimen. In the B scan, the time of flight (travel time) of the sound energy is displayed along the vertical axis and the linear position is displayed along the horizontal axis. The C scan representation provides a top view of the location and size of the test specimen features. C scan representations are usually produced with an automated data acquisition system.

In ultrasonic testing, the defect or flaw is identified by the amount of energy reflected from the defect. This amount is in turn dependent on the energy produced by the pulser. To ensure that the impedance of the pulser surface and the material surface match, so that maximum energy is incident for testing, a coupling agent is usually used. For all ultrasonic testing, a layer of water-based gel is applied before placing the transducer for efficient energy transfer.

Recently, a number of ultrasonic generating elements were put together in a single probe to make a phased array ultrasonic testing probe. The distinguishing feature of such probes is a computer control (amplitude and delay) of individual elements. The excitation of multiple piezocomposite elements can generate a focused ultrasonic beam with the possibility of dynamically modifying beam parameters such as angle, focal distance, and focal spot size through software. To generate a beam in phase by means of constructive interference, the various active transducer elements are pulsed at slightly different times, and similarly, the reflection from the desired focal point hits the various transducer elements with a computable time shift. The echoes received by each element are time shifted before being summed together. The resulting sum is an A scan that emphasizes response from the desired focal point and attenuates echoes from other points in the test piece.

Some of the common speeds of sound in various materials are given in Table 12.2.

A handheld portable phased array ultrasonic scanning system being used for internal defects is shown in Figure 12.2. An ultrasonic test is conducted to accurately determine the defect location in a weld in Figure 12.3 (a and b).

TABLE 12.2

Speed of Sound in Common Materials

Material	Sound Speed (m/s)
Aluminium	6260
Tin	3230
Iron	5900
Brass	4640
Copper	4700
Propylene resin	2730
Water	1480
Glycerine	1920

FIGURE 12.2
Handheld phased array ultrasonic probe during testing.

(a)

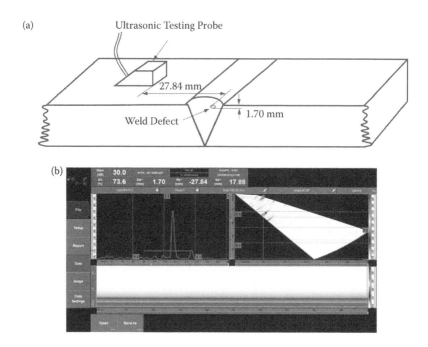

FIGURE 12.3
Ultrasonic weld testing: (a) location of weld defect, (b) sector scan by ultrasonic system.

12.4 Radiography

Radiography is used to determine internal defects by subjecting the surface to an X-ray. These are high-energy waves, and a density difference is due to a defect like a void or an intrusion. The transmitted beam intensity is inversely proportional to the density. By either obtaining a digital image or an image on X-ray-sensitive film, radiographs of the structure can be obtained. In tomography, the X-ray beam is rotated 360° and thus a three-dimensional image is obtained. Radiography is used to find scale formation inside boiler tubes, internal cracks, and so on.

12.5 Acoustic Emission

Materials under stress give out waves in the order of 2 MHz or higher. These waves can be sensed on the surface of a body through high-frequency piezoelectric-based acoustic emission sensors. Based on the intensity of the internal stress, these acoustic emission waves can be continuous or in bursts. By having multiple acoustic emission sensors on the surface of a body, and through high-speed data acquisition of the arrival times of these acoustic emission waves, the source of the internal defect can be determined through the method of triangulation.

13

Machine Tool Condition Monitoring

13.1 Introduction

Manufacturing is an important economic activity. It can be grossly divided into job-shop production, batch production, and mass production. Job-shop production delivers a very low volume of production with large variations in design. Batch production can provide some variation in design with low to moderate volume, whereas mass production caters to high volume with very low variety. Hard automation, for example in special purpose machine tools, transfer lines, and so on, has enabled economical large-volume production of engineering components. However, there has been a requirement of moderate variety in products with moderate volume. Neither batch production nor mass production can satisfy that requirement. The flexible manufacturing system (FMS) can economically provide such variation in design with moderate production volume. FMS utilizes the advantages of soft automation, computer numerical control (CNC) machine tools, and an automated and integrated manufacturing environment to deliver such products with enhanced product quality.

Lathes are one of the most common machining stations and are nowadays used with different cutting tools for different machining processes. Now, machining centers come with all the cutting tools arranged in a magazine from which the appropriate tools are used programmatically. The work piece can be placed with perfection on the order of micrometers. Part programs, written for CNC machining centers, control the whole process. These CNC machines are now being used from the manufacturing of very large scale integration (VLSI) wafers by micromachining to manufacturing of chassis of large space vehicles or missiles through blast machining and cryo-grinding.

There are a number of machining processes for different types of jobs and each of them has different working principles. Some of the common machining operations are shaping, turning, milling, drilling, grinding, and polishing, but the basic metal-removal strategy is the same: the work piece, or the cutting tool, or both rotate or move over one axis, and one of them is fed linearly to remove the metal from the exposed surface. The basic concept is shown in Figure 13.1 for a face milling process where the terms *cutting speed*, *cutting feed*, and *depth of cut* are defined.

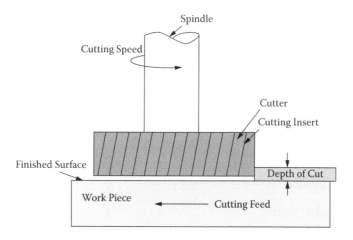

FIGURE 13.1
Face milling operation.

The state of tool wear affects the machining process and product quality. Therefore, it is important to observe and monitor the state of tool wear. There are several forms of tool wear such as flank wear, crater wear, notching, plastic deformation, chipping, and breakage. They can conveniently be categorized as regular or gradual and irregular phenomena. Irregular tool wear is unpredictable and may lead to a catastrophic failure, especially in brittle tools, which is undesirable in machining.

Tool wear in machining affects the following:

- Surface finish and surface integrity
- Dimensional accuracy
- Tool life
- Machine tool dynamics through vibration or chatter
- Power requirements

These factors not only affect the product quality but increase the cost as well. In order to produce components with the desired quality and to ensure smooth and efficient functioning of the machine tool, tool condition monitoring (TCM) is gaining acceptance in industries with the objective of enhancing productivity without sacrificing product quality and escalating cost.

During the process of machining, the softer work piece interact with the harder cutting tool, along with the removal of the work material, tool condition deteriorates. Due to this cutting-tool deterioration, the tool loses its desired shape for proper machining and hence the process is affected. The whole setup being designed with the proper machining mechanics in mind, the general dynamic performance of the system is affected.

Hence, the job and the cutting tool start to vibrate and the surface roughness increases. Moreover, in automatic machining in CNC machines, the cutting tool condition deteriorates cumulatively and hence aggravates the situation further. As the design of the tool insert is made with the smooth cutting operation in mind, mechanical forces acting on both the tool and work piece increase, and the stress limit of the tool may be reached. If the tool breaks while cutting, the force is released, and the shock is felt in both the machine tool and the job, both of which are costly. One way to avoid this may be to change the inserts in shorter periods. This increases the cost of manufacturing significantly. Moreover, with phenomena like microchipping and attrition in inserts being highly unpredictable, catastrophes cannot be eliminated.

Hence, product quality is hampered and cost goes up as well. In the present competitive market, quality and cost are both crucial factors, and there must be a system to detect tool wear before the tool breaks on the job; maybe even long before that, so that surface finish and quality can be maintained. The solution is continuous tool condition monitoring (TCM). If the condition is found to be alarming, proper decision can be taken, the last resort being rejection of the old inserts and fixing fresh inserts.

Tool condition means the overall state of the cutting tool and it is a qualitative measure. Even if it is quantified with a numerical scale, it is arguable whether that can be used in all domains of the manufacturing industries. The main cutting-edge wear in a cutting tool is a very good indicator of the tool's condition, and being a numerical value, is acceptable to any industry based on its uniqueness.

13.2 Tool Wear

In all machining operations a cutting tool is used. The cutting tool has sharp edges that are responsible for material removal due to relative motion between the cutting edge and the work piece. Milling is an important material-removal process in automated manufacturing. The cutting tool wear phenomenon with respect to milling is described here. A face milling process is an intermittent machining operation. The inherent problems in the face milling process are mechanical shocks at the entry and exit of the cutting tool, dynamic loading due to varying chip thickness, thermal cycling, foot formation, and vibration and noise during metal removal. The degree of such problems increases further with small entry and exit angles, very high speed and feed rates, high strength and hardness of the work material, unfavorable machining environment, and insufficient rigidity of the machine tool. All these problems impair the performance of carbide cutting tools, which are brittle in nature. Moreover, these problems cause chipping, cracking, and fracturing in tool tips, in addition to usual gradual wear at the face and flank of the inserts, which causes premature and random failure in the cutting tools.

For high material removal rates (MRRs), carbide inserts are generally used in face milling operation. Single carbide is used for brittle material like cast iron, and composite carbide for ductile material like steels. The cutting edges of a tool are shown in Figure 13.2 and the same when viewed under a scanning electron microscope in Figure 13.3.

The main cutting edge removes the major portion of the work material, the planishing edge is provided for good surface finish by removing the feed marks, and the small intermediate edge is provided for tip strength and reduction and favorable distribution of temperature at the tool tip. In machining, the chip moves over the rake surface of the cutting tool with high velocity and the contact between the chip and tool is subjected to high machining temperature at sufficiently high contact stress. Similarly, the finished machined surface rubs against the principal and auxiliary flank

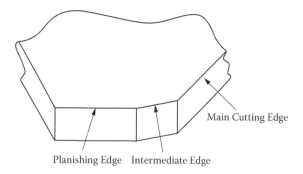

Planishing Edge   Intermediate Edge

FIGURE 13.2
Edges of a cutting tool.

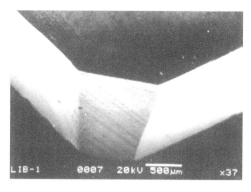

FIGURE 13.3
Scanning electron microscope (SEM) view of a fresh cutting tool.

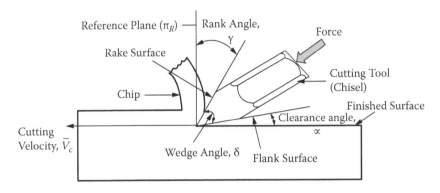

FIGURE 13.4
Wedge, rake, and clearance angle of cutting tools.

of the cutting tool. Such rubbing action associated with high contact stress and temperature leads to tool wear. The terminology and cutting surfaces of a single-point cutting tool and its geometry are schematically shown in Figure 13.4.

13.3 Sensor Fusion in TCM

For online TCM, it is not possible to measure a direct signal to characterize tool condition. We need to find some other signal, which is affected by the tool condition, so that the signal can be used as an indicator of tool condition. A one-to-one relationship between the tool condition and the selected signal would be ideal, but unfortunately, no signal has been reported that has the unique quality of being affected by tool condition alone, although some signals are found to be affected by many factors including tool condition. Somehow, if the effects of the factors other than the tool condition are eliminated, these signals can be used to estimate the tool condition.

Elimination of the effect of factors other than the tool condition on the signal is a difficult task. Specifically, signals like the cutting force signal from a dynamometer, vibration signal from an accelerometer, and so on, which are reported to be good indicators of the tool condition, are also found to be affected significantly by the work material, machine tool rigidity, machining process and process parameters, cutting condition, and so on. So a single signal cannot directly provide the desired information for indirect online estimation of the tool condition.

Information integration and sensor fusion is one technique that can be used to solve this problem. Candidate signals like force, vibration, power, sound, and acoustic emission are dependent on different unwarranted factors, and fusing them into an estimate can give the desired information on tool condition.

13.4 Sensors for Tool Condition Monitoring

There are mainly two approaches in tool condition monitoring—offline and online. Further, TCM can employ a direct or indirect approach depending on measurement techniques. In direct TCM, the state of tool wear is directly assessed. For example, the image of the tool may be grabbed by a charge-coupled device (CCD) camera and analyzed to estimate tool wear. In the indirect method, the tool wear is estimated by monitoring some related machining characteristic like cutting force or vibration.

13.4.1 Direct Tool Wear Measurements

Generally, for online TCM, the indirect approach is preferred, since measurement of tool wear on the fly is not convenient, though there have been attempts made to have an online vision-based system to measure tool wear in real time. Some of the commonly used direct tool wear measurement systems are discussed in the following sections.

13.4.1.1 Dimensional Deviation

Deviation in the dimension of the work piece is measured using any of the contact or noncontact transducers, as dimensional deviation of the work piece increases with cutting tool wear with the same operational settings. With the same concept, a noncontact-type laser beam can be used to measure the work piece diameter during machining. It has been employed in the turning process and detected the dimensional deviation including out of roundness. An optoelectronic device can also be used for this concept.

However, the limitation of such systems is that it cannot distinguish between the various tool failure modes. Moreover, dimensional deviation may arise from other sources of error such as the tool fixture system, misalignment and faulty movement of the slides, and so on.

13.4.1.2 Tool-Work Electric Resistance and Radioactivity Analysis

Electrical resistance between the job and tool decreases with the progression of tool wear. The limitation of this process is distortion due to temperature, humidity, force, and plastic deformation. In radioactive techniques, the cutting tool edge has to be treated with radioactive material. The under surface of the chip will pick up radioactivity from the tool, and radioactivity from the tool and radioactivity of the chips can be related to volumetric tool wear. The hazards of radiation is the limitation of this technique.

13.4.1.3 Optical Sensors

Optical sensors and cameras can be used to analyze cutting tool wear. The reflectivity of worn tool surfaces is greater than the unworn tool surface. Laser beams and digital cameras are used for morphological detection of different tool wear like notch wear, crater wear, and chipping, but a minimum of 0.05 mm/pixel resolution is required for measuring different tool wear. Thus, the resolution of the camera greatly affects the accuracy of the tool wear measurements. In addition to the sensor, a frame grabber, powerful lights, and processing units are necessary. But in a commercial camera, the typical field of view for 4- to 6-mm depth-of-cut machining can be obtained. Further, for the optical sensor, the tool has to be withdrawn from machining, which affects productivity.

13.4.2 Indirect Tool Wear Measurement

On the other hand, in online TCM, an enormous number of sensors are available for indirect measurement of cutting tool condition. Some of them are described in the following sections.

13.4.2.1 Force Sensor

The force sensor heads the list because the magnitude of variation of forces with flank wear is significant. Cutting forces can be measured with different kinds of force sensors, like direct measurement dynamometers, plates and rings, pins and extension sensors, and force measuring bearings. Because of close sensitivity to flank wear, these force sensors are used in online TCM. A work piece attached on top of a cutting force dynamometer is shown in Figure 13.5.

13.4.2.2 Vibration Sensor

The vibration level of machining process varies with cutting tool condition. The vibration primarily concerns the amplitude, frequency, and acceleration of work-holding or tool-holding devices in one or more directions. This sensor used for online measuring of vibration level relates to rubbing of the tool flank and work, formation of built-up edges, and cutting condition. A piezoelectric triaxial accelerometer attached to the base plate of a work piece being surface milled is shown in Figure 13.5.

FIGURE 13.5
Force dynamometer and triaxial accelerometer installed during face milling.

13.4.2.3 Surface Roughness

Under a given condition, the roughness of the machined surface is influenced by the condition of the cutting edges. In online TCM, laser beams are used to measure surface roughness. The reflection pattern of the laser beam correlates very well with the surface roughness of the machined job, but the need of cleaning the machined surface to remove oil, grease, and coolants used during the machining process hinders commercial implementation.

13.4.2.4 Cutting Temperature

During the machining process, heat is generated at the cutting zone due to primary and secondary deformation of the work material, and rubbing of the tool with the chip and work piece. Development of tool wear causes more friction between the tool and work, leading to changes in machining temperature. Therefore, temperature is another characteristic that can be used in a TCM system. The average temperature in machining is generally measured by thermocouples that detect the variation of electromotive force (emf) generated at the hot junction of the work–tool interface. Limitation of this system is that the generated voltage signal may be distorted by a parasitic emf, and as such, calibration becomes faulty.

Temperature can also be measured by the thermal radiation coming from the tool surface with the help of a pyrometer or infrared photography. The drawback of this method is the lack of close access to the cutting zone and poor response to different tool failure modes, which reduces the effectiveness of temperature signals in TCM.

13.4.2.5 Spindle Current, Voltage, and Power

The amount of torque required to remove the work material increases with increase in tool wear, and so does power. Monitoring the spindle current and voltage, the power required can be determined. Power has been measured to detect cutting tool wear in turning, drilling, and milling processes.

13.4.2.6 Torque Sensor

Normally, spindle current and torque are used to measure power during the metal cutting process. Both can be used to determine power requirements in the cutting process and provide information about the dynamics of cutting. Determination of power by measurement of torque is more sensitive, since the torque sensor is located close to the cutting tool. However, measuring torque is more complicated than measuring the current of a spindle motor. So in many cases, the spindle current and feed drive current are measured because the spindle current corresponds to torque and the feed drive current corresponds to thrust force. This feed drive current measurement is a good indicator of tool wear and tool failure.

13.4.2.7 Microphone

Sound is generated because of rubbing of the tool on the work piece when the tool undergoes wear and chipping. The low-frequency components of the machining sound are analyzed to estimate the tool wear; the cutting pressure increases with increased flank wear and reaches a point of inflection, after which the intensity level decreases sharply. A microphone may not be suitable in an industrial environment because of high background noise.

13.4.2.8 Acoustic Emission Sensor

Acoustic emission (AE) is transient elastic stress energy released spontaneously when a material undergoes deformation. AE signals are two types in nature—continuous AE and burst AE signals. Continuous AE signals appear as a result of yielding and burst AE signals are generated due to chipping and breakage of the tool. Hence, this is mainly used to predict fracture and chipping before they happen. This has been extensively experimented upon in turning, drilling, and milling. AE signals are generated not only by breakage or chipping in the cutting edge, but also rubbing action on the work surface by the unbroken chips and chip-formation process. AE signals

FIGURE 13.6
AE sensor near the spindle.

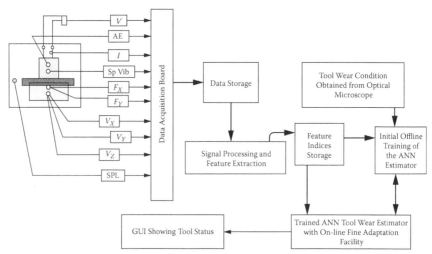

FIGURE 13.7
System architecture for tool condition monitoring.

may even increase or decrease depending upon the shape of the fractured tool. An AE sensor mounted close to a milling machine spindle is shown in Figure 13.6.

13.5 A Tool Condition Monitoring System

The objective of online tool condition monitoring is to estimate the wear status of the cutting tool in operation. A general system architecture for a tool condition monitoring system is shown in Figure 13.7. This system essentially consists of a machining system on which various sensors are installed with an objective of collecting the system's response during the machining process. Since no single sensor is best suited for online tool wear monitoring, an option of multiple sensors is chosen. The signals from the sensors are acquired using a multichannel data acquisition system. The digitized data are stored in a database. Digital signal processing is done on the signals to extract the signal features. These features in the time domain are parameters like mean, root mean square (RMS) value, kurtosis, skewness, and so on. Frequency domain features of the signals can also be used. Many times, digital filtering is done to remove artifacts in the signal, which may be redundant and not useful. Many models based on probability and statistical learning theory are available for tool condition estimation. Unlike rotor-bearing systems, where using vibration signature analysis one can diagnose the fault in a machine, in a machining system, using vibration measurements alone has not led to an accurate estimation of tool wear, since in a machining center there are several vibration sources at the same time, giving rise to signals with poor signal-to-noise ratios. Self-computing technique like artificial neural networks (ANN) in that sense has been popular since once in an offline mode, if the actual tool wear can be measured using a microscope and a multi-input plant transfer function can be developed using the data from the multiple sensors, a robust prediction of the tool wear can be made from the trained plant transfer function. In the system architecture shown in Figure 13.7, the ANN modules for training and wear estimation are shown. Commercial ANN-based embedded systems are in place in CNC machining centers, which provide online indications of tool wear condition and provide alarms when replacement is due, though other techniques like support vector machines (SVMs), fuzzy logic, genetic algorithms, and so on are also popular for fault diagnosis and prognosis in manufacturing systems.

13.5.1 Tool Wear Estimation in a Face Milling Operation

Table 13.1 gives the machining condition for a typical face milling operation where a tool condition monitoring system is in place. Several types of transducers to measure cutting force, spindle vibration, and spindle motor current were

TABLE 13.1

Face Milling Conditions

Machine Tool	CNC Plano-Miller
Work material (size)	C-60 Steel (150 mm × 70 mm × 70 mm)
Cutting insert	WIDIA SPKN 1203 EDR TTMS
Cutting condition	Dry
Cutting velocity (V_c)	98 m/min, 110 m/min, 140 m/min, 212 m/min
Cutting feed (S_0)	0.08 mm/tooth, 0.16 mm/tooth, 0.24 mm/tooth
Depth of cut (t)	1 mm, 1.5 mm, 2 mm
Cutting insert	Single insert

installed on the machine. Figures 13.8 and 13.9 show the scanning electron microscope (SEM) views of the worn and chipped cutting tool, respectively. The measured parameters of the cutting force and motor current increase with the tool wear as shown in Figures 13.10 and 13.11, respectively. Usually, a tool is replaced if its wear increases beyond 500 μm. Such curves can also be linearly extrapolated to indirectly determine the tool wear condition.

FIGURE 13.8
SEM view of worn cutting tool showing all the three edges.

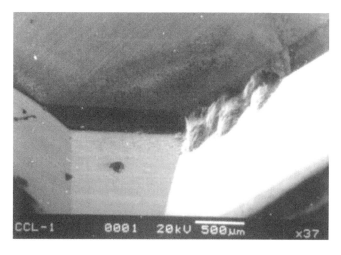

FIGURE 13.9
SEM of chipped cutting tool.

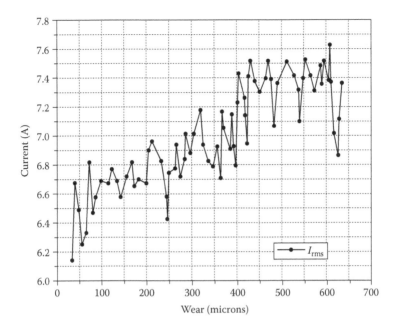

FIGURE 13.10
Variation of spindle motor current with tool wear.

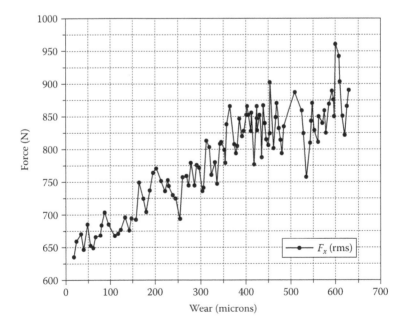

FIGURE 13.11
Variation of cutting force with tool wear.

FIGURE 13.12
CNC turning center.

13.6 Other Manufacturing Operations

Turning and drilling are also important machining operations that are done by CNC machines. TCM that is similar to the TCM described in earlier sections can also be done in such machines or in intelligent manufacturing systems. A CNC-based turning center where online TCM can be implemented is shown in Figure 13.12.

14

Engineering Failure Analysis

14.1 Introduction

Despite the best maintenance, a machine may fail. The reasons for failure are many, like inherent design flaws, improper material choice or material defects, manufacturing or installation defects, and service life anomalies. In such a situation, the diagnostician is usually left to perform a root-cause analysis of a failure. In this chapter, an overview and step-by-step procedure for conducting a failure analysis of failed machine components is presented.

14.2 Overview of Failure Analysis

A general overview of the steps to be followed for failure analysis is presented here. Documentation and past history are very important in providing a lead to the possible cause of the failure. The condition-based maintenance (CBM) automation systems in place usually archive the data in software databases from sensors around the machines. At any given time, the past data collected from the machine can be retrieved. In machineries, the operating conditions in terms of loads and rotational speed are helpful in estimating the stress developed at the failed or critical components of a machine. The original design specifications of the machine usually provide such data. At the failed site, a good photographic record helps the investigator in reconstructing the possible failure modes of the machinery. Collection of failed components from the test site for further nondestructive tests (NDT) and metallurgical analysis is to be done. Once all the tests and analysis have been done, a possible conclusion can be drawn regarding the reason behind the failure.

14.3 Failure Modes

Fractured or failed surfaces show telltale signs of the mode of failure of metal machine components (Figure 14.1). Table 14.1 provides a description of the fractured surfaces for different types of failure modes.

14.4 Failure Analysis

The primary reasons for conducting an analysis of a metallurgical failure are to determine and describe the factors responsible for the failure of the component or structure. From an engineering standpoint, the proper application of failure analysis techniques can provide valuable feedback on design problems, material limitations, and manufacturing flaws.

The optimum design is one in which the requirements are slightly exceeded by the capabilities in all circumstances. This aim is seldom realized because

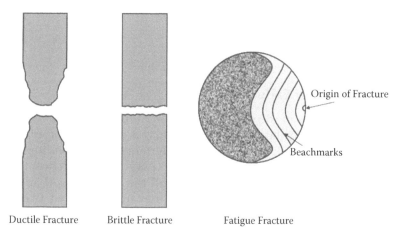

Ductile Fracture   Brittle Fracture   Fatigue Fracture

FIGURE 14.1
Views for different modes of fracture: ductile, brittle, and fatigue.

TABLE 14.1

Fracture Mechanisms and Their Surface Characteristics

Mode of Fracture	Typical Fracture Surface Characteristics
Ductile	Cup and cone, dimples, dull surface, inclusion at the bottom of the dimple
Brittle	Shiny, grain boundary cracking
Fatigue	Beachmarks, striations, propagation area, zone of final fracture

of the obvious difficulty in recognizing or defining precisely the various demands that the system will be called upon to meet.

This latter aspect of the design requirements is generally met by a sound engineering practice, the application of safety factors.

However, how much of a safety factor is appropriate? To grossly overdesign the component is economically extravagant and can inadvertently overload other parts of the structure. Underdesigning the component leads to its premature failure, is economically wasteful, and most important, could jeopardize life.

14.4.1 Manufacturing and Installation Defects

14.4.1.1 Metal Removal Processes

- Cracks due to abusive machining
- Chatter or cracking due to speeds and feeds
- Microstructural damage due to dull tool
- Grinding burn
- Electrical discharge machining recast layer cracking
- Electrochemical machining intergranular attack
- Residual stress cracking due to overheating

14.4.1.2 Metal-Working Processes

- Cracking, tears, or necking due to forming/deep drawing
- Laps due to thread rolling/spinning
- Tool marks and scratches from forming
- Surface tears due to poor surface preparation prior to working
- Residual stress cracking due to flow forming
- Lüders line due to forming strain rate
- Microstructural damage due to shearing blanking piercing
- Overheating damage during spring winding
- Laps and cracks due to shot peering
- Stress-corrosion cracking due to use of improper die lubricants

14.4.1.3 Heat Treatment

- Grain growth
- Incomplete phase transformation
- Quench cracks
- Decarburization
- Untempered martensite

- Temper embrittlement and similar embrittlement conditions
- Inadequate precipitation
- Sensitized microstructure
- Inhomogeneities in microstructure
- Loss of properties due to overheating during postplating bake

14.4.1.4 Welding

- Lack of fusion
- Brittle cracking in heat-affected zone (HAZ)
- Sensitized HAZ
- Residual stress cracking
- Slag inclusions
- Cratering of fusion zone at endpoint
- Filler metal contour out of specification
- Hot cracking
- Cracking at low exposure temperatures
- Hydrogen embrittlement due to moisture contamination

14.4.1.5 Cleaning/Finishing

- Corrosion due to inadequate cleaning prior to painting
- Intergranular attack or hydrogen embrittlement due to acid cleaning
- Hydrogen embrittlement due to plating
- Stress corrosion from caustic autoclave core leaching of castings

14.4.2 Assembly at Factory/Installation at Site

- Misalignment
- Missing/wrong parts
- Improper fit-up
- Inappropriate fastening system, improper torque
- Improper tools
- Inappropriate modification
- Inadequate surface preparation

14.4.2.1 Inspection Techniques

- Arc burn due to magnetic particle inspection
- Intergranular attack or hydrogen embrittlement due to macrotech
- Fatigue or quench crack from steel stamp mark

14.4.3 Laboratory Analysis

- Initial examination
- Photodocumentation
- Nondestructive destruction
- Material verification
- Fractographic examination
- Metallurgical analysis
- Mechanical properties determination
- Analysis of evidence
- Writing of a report

14.4.4 Material Selection

- Tensile strength
- Yield strength
- Modulus of elasticity
- Ductility (percent elongation)
- Fatigue strength
- Fracture toughness
- Hardness
- Shear strength
- Machinability
- Coefficient of friction
- Impact strength
- Corrosivity
- Density
- Coefficient of thermal expansion
- Thermal conductivity
- Electrical resistivity
- Other physical properties

14.4.5 Failure Investigation Procedure

- Documentary Evidence
- Service Conditions
- Materials Handling, Storage, and Identification
- Interviews
- Testing

14.5 Failure Analysis Sampling Guide

The best results in failure analysis are obtained when residue can be evaluated without spoilation from repairs and poor sampling processes. The following is a basic guide for selection of samples for failure analysis and recommendations for sample removal with minimum impact on subsequent evaluation.

14.5.1 Before Beginning Sample Removal

1. Thoroughly examine the failed component, surrounding areas, and all mating components to document the failure characteristics and other notable conditions.
2. Document using digital photography, if possible. Photograph from several views and angles for subsequent verification of details that may not be recorded in visual inspection notes. Use a ruler to scale the failed component in photograph.
3. Record any operating condition data from the failure that may be available from systems monitoring equipment.
4. Interview operators and other witnesses for descriptions of the failure events and circumstances.

14.5.2 Selection of Samples for Laboratory Evaluation

1. If possible, remove the entire failure region with some adjacent sound material.
2. If failure sites are numerous, try to select samples of failure sites with different appearances and/or from locations with potential differences in operating conditions.
3. In many cases, it is useful to remove mating components, as well as the failed parts, e.g., collect the nuts from failed fasteners or gears and bearings with a failed shaft.
4. For large structures, obtain material sample(s) well away from the failure to allow comparison of material properties, e.g., sample a ruptured boiler tube at the failure and in a colder regions such as near the tube sheet.

14.5.3 Sample Removal

1. Make any required cuts well away from the failure to avoid mechanical deformation and other damage that can obscure evidence at the failure site.
2. Whenever possible, make cuts with a toothed saw and no lubricant.

3. If flame cutting or abrasive cutting without lubricants must be done, make cuts far enough away from the failure site to avoid heating the material at the failure. Allow at least 3 inches for abrasive cuts and 6 inches for flame cuts, but actual distance depends on the size of the section.

4. Avoid using coolants when removing corrosion failures. Coolants will contaminate the site and obscure evidence of the corroding medium.

14.5.4 Packing and Shipping

1. Handle removed samples gently. Avoid touching failure sites with bare hands and do not place mating fracture surfaces together. Even light contact can smear microscopic fracture features or dislodge fragile corrosion products.

2. If samples are wet, dry in a gentle air stream before packing. In most cases, it is preferable that samples are not coated with any protective oil or other product before packing.

3. Pack samples in sealed bags or wrapped tightly in plastic. Provide cushion between samples and outside of container to avoid damage in transit.

14.5.5 Assembling Background Data

The investigator must develop a complete case history of the component, including details of the failure and information about the manufacturing history before he/she can intelligently select those tests and procedures that are best used to analyze the failure. Important items to be determined include the following:

14.5.5.1 Information about the Failed Components

- Location, name of item, identifying numbers, owner, user, manufacturer, and fabricator
- Function of item
- Service life at the time of failure
- Rating of item, operating levels, normal and abnormal loads, frequency of loading, and environment
- Material used
- Manufacturing and fabrication techniques used, including specifications and codes governing the manufacturing, fabrication, inspection, and operation of the component
- Normal stress orientation, operating temperature range, pressure. and speeds

- Strength and toughness
- Heat treatment, stress relief, or other thermal processing
- Fabrication procedures, such as welding, adhesive joining, coatings, bolting, and riveting
- Inspection techniques and quality control reports during manufacturing, service and maintenance records

14.5.5.2 Information about the Failure

- Date and time of failure, temperature, and environment
- Extent of damage, sequence of failure and injuries
- Stage of operation when failure occurred
- Blueprints, photographs, or sketches of the failure and adjacent areas
- Any service deviations that might have contributed to the failure
- Opinions of operating personnel regarding the failure

14.5.6 Analyzing the Data

While performing the analysis, the investigator should cross check each observation against the history of the part and note any contradictions. It is frequently just such contradictions that reveal a possible cause of failure. In evaluating these factors and assessing their significance, the investigator should draw not only on his experience but also on pertinent disciplines other than his own special field.

After completion of the previously mentioned procedures, the metallurgist is ready to interpret and summarize the facts that have been gathered. The absence of certain features is often as important as the presence of these factors. Most failures are caused by more than one factor, although frequently one factor may predominate. Again, the metallurgist should make use of his own knowledge and experience as well as that of others in establishing and rating the importance of these factors. Occasionally, given the nature of the data or the theoretical state of knowledge about the particular causative mechanisms involved, the final conclusion may simply be a matter of the investigator's best judgment as to the most important factor and should be so stated in the report.

Examination of the failure and a study of the various conditions pertaining to the design, operation, and environment should raise certain questions:

- What was the sequence of the failure? The speed? The path?
- Were there one or more initiation sites?
- Did the failure initiate at or below the surface?
- Was the failure located by a stress concentrator?
- How long was the crack present?

- What was the intensity of the load?
- Was the loading static, cyclic, or intermittent?
- How were the stresses oriented?
- What was the failure mechanism?
- What was the approximate temperature at the time of failure?
- Was the temperature important?
- Was wear a factor?
- Was a corrosion a factor? What type of attack?
- Was the proper material used? Is a better material required?
- Was the cross section sufficient?
- Was the material quality acceptable for the grade and specifications?
- Was the component properly heat treated?
- Was the component properly fabricated?
- Was the component properly assembled, aligned, and so on?
- Was the component repaired during service properly?
- Was the component properly maintained? Lubricated?
- Was the failure caused by a service abuse?
- Can the design be improved to prevent similar failures?

14.5.7 Preparing the Failure Report

The report analyzing the failure should be written in a clear, concise, logical manner. It should be clearly structured with sections covering the following:

- Description of the failed item
- Conditions at the time of failure
- Background history important to the failure
- Mechanical and metallurgical study of the failure
- Evaluation of the quality of the material
- Discussion of any anomalies
- Discussion of the mechanism or possible mechanisms that caused the failure
- Recommendations for the prevention of future failures or for action to be taken with similar pieces of equipment

14.5.8 Preservation of Evidence

Due to the rise in number of litigations and the strict implementation of court laws, it is necessary to properly preserve evidence.

15

Case Studies

15.1 Introduction

This chapter presents a few cases of machinery troubleshooting and root cause of failure of different types of machines encountered in the field. The case study examples given here provide an overall insight into how vibration measurement and analysis, stress analysis, motor current signature analysis, and fractography aid the machinery diagnostician in determining the root cause of failure. The examples given here have been encountered by the author.

   i. Failure analysis of a raw material handling conveyor bend pulley at a cargo shipping port

   ii. Failure analysis of a cement plant

   iii. Failure and strength estimate of a conveyor support structure

   iv. Vibration measurements on a motor-multistage gearbox drive system

15.2 Bend Pulley Failure Analysis

A shipping port company, handling raw material for steel plants was concerned with the repeated failures of the bend pulley in the conveyor system at the port. They were interested in determining the root cause of the failure and suggest remedial measures involved the following:

1. Perform a design audit for loads on the bend pulley

2. Determine the resonance conditions

3. Determine the material composition

   In order to carry out the work, an on-site visit was undertaken. Material samples from the failed pulley were collected, the shell of the failed pulley was visually inspected after the lagging was removed, and operation of the conveyor drive system was also observed.

In order to check the design and resonance conditions of the bend pulley, finite element analysis was performed on the bend pulley, with the load data and drawings provided by the client. The material composition was measured at a test laboratory using an atomic absorption spectrophotometer. The microstructure of the samples of the bend pulley were also studied under various magnification conditions. The results of the study are presented in the following sections, along with the conclusions and recommendations for avoiding such failures in the future.

15.2.1 Design Audit

Figure 15.1 shows the loads on the failed bend pulley as provided by the client. A stress analysis and natural frequency estimation using finite element analysis of the bend pulley was done using a commercial finite element modeling (FEM) software. The FE model of the failed pulley is shown in Figure 15.2.

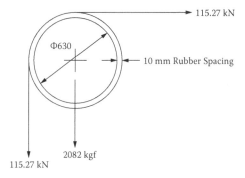

FIGURE 15.1
Loads on the failed bend pulley.

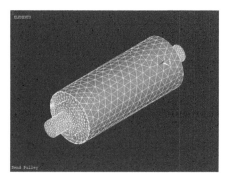

FIGURE 15.2
FE model of the failed bend pulley.

The material for the shell was Fe410WB as per IS:2062 with a yield point strength of 285 MPa, ultimate tensile strength (UTS) of 468 MPa, and percentage elongation of 29.44%. The shell is made of hot rolled steel plate of the same material.

A static stress analysis on the failed pulley as per the load given in Figure 15.1 was made. The stress and deflection in the pulley are given in Figures 15.3 and 15.4.

The maximum deflection in the pulley is 7.68 mm, which occurs in the center of the pulley.

The maximum von Misses stress developed in the pulley is 8.07 MPa, which is very low compared to the allowable maximum yield point stress of the material. This maximum stress occurs at the bearing locations of the shaft. There is no appreciable stress at the end plate and shell weld connection region.

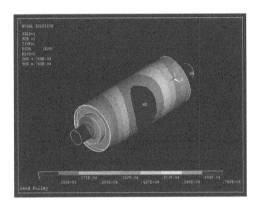

FIGURE 15.3
Deflection of the failed pulley.

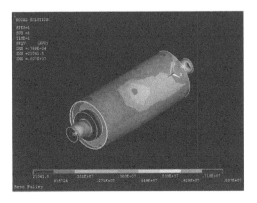

FIGURE 15.4
Von Misses stress developed in the pulley.

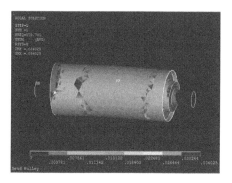

FIGURE 15.5
First mode shape of the pulley at 170 Hz.

15.2.2 Resonance Conditions

Since the conveyor has a linear speed of 3.8 m/s with a material flow rate of 5500 ton/hour, the rotational speed of the shell having a diameter of 630 mm, is calculated to be 115 RPM [1.9 Hz]. The rotational forcing frequency is quite low and on the order of only 2 Hz. The natural frequency of the bend pulley during rotation about its axis was estimated be around 170 Hz, as shown in Figure 15.5. This frequency is way above the forcing rotational frequency, and hence the condition of resonance does not occur during the pulley's operation.

The subsequent natural frequencies of the pulley are beyond 170 Hz, and the fourth natural frequency is at 380 Hz.

The above static and dynamic analysis indicate that the pulley is safe from a strength point of view during its operation. However, since the pulley has failed, focus in the investigation was shifted to the metallurgy of the pulley material and the manufacturing aspects of the same.

15.2.3 Metallurgical Composition

The chemical composition of the shell material Fe410WB is given in Table 15.1. The collected sample of the shell material was measured by an atomic absorption spectrophotometer.

The composition of the shell is not significantly different than that specified by the original equipment manufacturer (OEM), though a higher percentage of manganese is observed.

15.2.4 Hardness

The OEM recommended hardness of the shell material is around 120–160 BHN. The measured hardness of the failed shell sample close to the circumferential weld is around 45.2 HRA, which corresponds to a very high hardness. This leads to a conclusion that the microstructure of the failed sample

TABLE 15.1

Chemical Composition of the Shell Material

Chemical Element	Percentage Composition by OEM	Measured Composition of the Failed Sample
Carbon	0.180	0.184
Manganese	0.780	0.924
Phosphorous	0.028	0.047
Sulphur	0.030	0.016
Silicon	0.260	0.121
Chromium	—	0.003

needs to be looked into, because the failed component along the edge of the shell has a brittle mode of failure, as can be seen in Figure 15.6.

15.2.5 Visual Inspection of the Failed Shell

Initially it was reported that the bend pulley shell had failed close to one of the end plates, as shown in Figure 15.6. However, during the on-site inspections, once the lagging on the shell material was cut and removed, a series of longitudinal cracks along the circumference of the shell was noticed, as shown in Figures 15.7 to 15.9.

It is observed from the corrosion marks near the cracks, that the initial cause of failure is the initiation, generation, and propagation of these cracks along the length of the shell, from one end of the shell, which weakened the shaft, and then the region near the heat-affected zone (HAZ) of the circumferential weld gave way, leading to a catastrophic brittle mode of failure as shown in Figures 15.10 and 15.11. This is substantiated by the microstructure study of the failed material given in the next section.

15.2.6 Microstructure Analysis

The microstructure of the failed shell samples were studied under the microscope and Figures 15.12 and 15.13 show the microstructure of the sample under various magnifications.

The presence of ferrite and pearlite indicates that austentite was present in the hot rolled steel plate from which the bend pulley was manufactured by plate bending and welding. It is understood from the manufacturer that CO_2 gas-submerged metal arc welding (GMAW) was used. The PWHT (post-weld heat treatment) significantly influences the mechanical properties. The HAZ is where the mechanical properties and microstructure are altered during the welding process. However, in this case the parent plate (hot rolled) is significantly anisotropic and care has to be taken in the PWHT. This anisotropy is because of the banded presence of ferrite and pearlite.

(a)

(b)

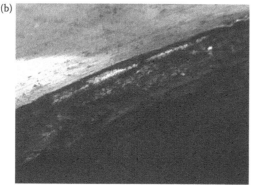

FIGURE 15.6
Brittle failure of the shell near the weld to the endplate: (a) view of the failed bend pulley, (b) view of the failed edge.

This banded microstructure affects the ductility and impact energy of the steel. High levels of manganese are also responsible for this banding, and austenitizing temperature, austenite grain size, and cooling rate during hot rolling influence the severity of the microstructural banding. The banding gives rise to alternate zones of hard and soft material, and during the fatigue loading of the shell, the cracks initiate from these areas and with time grow toward the edge of the shell.

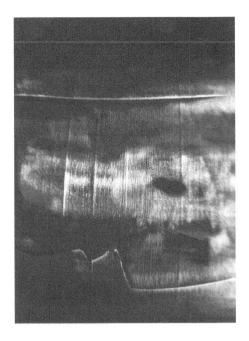

FIGURE 15.7
Longitudinal crack on the shell underneath the lagging.

FIGURE 15.8
Longitudinal crack propagating and ending at the end plate.

FIGURE 15.9
Corrosion of the cracked edges beneath the lagging.

FIGURE 15.10
Cracks in three directions beneath the lagging.

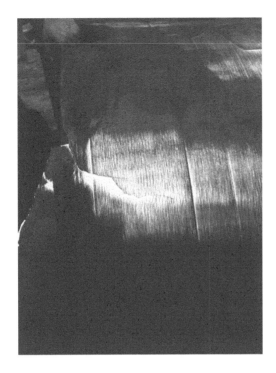

FIGURE 15.11
Edge crack near the weld zone.

FIGURE 15.12
Microstructure under 200X magnification, showing banded ferrite–pearlite microstructure of the failed shell steel.

FIGURE 15.13
Above structure under 2000X magnification, light (ferrite) and dark (pearlite).

15.2.7 Summary of Observations

From the above studies of the failed shell, the following conclusions are made:

1. The static and dynamic design of the bend pulley is within safe limits for operation.
2. The material composition of the shell as specified by the OEM is permissible and acceptable.
3. The in situ hardness at the failed area of the shell has increased.
4. The cracks were initiated/formed in the shell during the thermal cycle of weld and fatigue loading on the shell.
5. The banded nature of the microstructure has contributed to the early failure of the shell.
6. The hot rolled steel plate used to fabricate the shell did not have uniform grain size, and was cooled at a faster rate to provide banded structure.
7. Though the OEM claims to have conducted the welding per ASME section IX, special care is to be taken while welding steel plates with banded microstructures.

15.2.8 Recommendations

1. Proper choice of the hot rolled steel plate is to be made ensuring that it does not have a banded microstructure. Hot rolled plates with uniform grain size are to be selected for manufacturing the shell.

2. Special care is to be taken in the PWHT; for instance, the weld temperature has to be held for about an hour in the case of 20-mm steel plate. Extreme slow cooling is to be done.

3. Early detection techniques for cracks in the shell should be put in place at the critical bearing locations, by performing vibration signature–based analysis. The bearings should be self-aligning like the present ones.

4. In situ balancing of the bend pulley should be done.

5. Provisions of periodic visual inspection and ultrasonic testing for early crack detection should be put in place at the plant.

15.3 Root Cause Analysis of Torsion Shaft Failure in a Cement Plant

In a cement plant, a major concern was the excessive vibration in the plant during a drop in the electrical supply phase voltage to the electric motor driving the gearbox that rotated the cement kiln at a very slow speed of 3 RPM. Through a detailed vibration measurement and analysis at critical points on the gearbox, the possible cause of the plant vibration is determined and remedial measures suggested.

15.3.1 Vibration Measurement

The vibration was measured at six locations in the plant in all three directions (axial, vertical, and horizontal), as shown in the Figures 15.14 through 15.18, while the plant was in normal operation and when there was excessive

FIGURE 15.14
Location 1 at the DE bearing of the electric motor.

FIGURE 15.15
Location 2 on the input shaft of the gearbox; also note the photo-tacho for rotational speed measurements.

FIGURE 15.16
Location 3 on the output shaft of the gearbox driving the torsion shaft.

vibration. During the entire duration of measurements, the rotational speed of the motor output shaft was recorded and monitored online through a photo-tachometer. The measured vibration levels are given in Table 15.2.

During the course of measurements, it was observed that suddenly the plant vibration increased and a lot of hammering impact on the floor of the plant was felt. The speed measurements were monitored and were seen to fluctuate from 16.25 Hz (975 RPM) to 16.62 Hz (997.2 RPM). The phase voltage to the motor that was being recorded was hovering around the range from 6100 V to 6300 V.

The vibrations during the abnormal vibration regime of the plant are given in Table 15.3, where XX indicates that the levels were too high to measure.

15.3.2 Vibration Analysis

The recorded vibration and speed were analyzed in a multichannel fast Fourier transform (FFT) analyzer. It was observed that during the normal

FIGURE 15.17
Locations 4 and 5 on the drive end bearing block of the pinion.

FIGURE 15.18
Location 6 on the nondrive end bearing block of the pinion.

operation of the plant the rotational speed had small fluctuations around 16.54 Hz, however, when the voltage supply to the motor dropped, the speed came down to about 16.25 Hz, and in this condition the speed also came up to 16.62 Hz. Thus, there is a strong torsional acceleration being set up in the motor, and the corresponding fluctuating torque is responsible for setting up strong torsional oscillations in the gearbox, torsion shaft, and the pinion. Since the torsion shaft is the lightest of all the downstream components, the amplitude of vibration oscillations are also high.

TABLE 15.2

Vibration in m/s² at the Plant during Normal Operation,
Motor Speed around 16.54 Hz (992.4 RPM)

Location	Axial	Vertical	Horizontal
1	1.37	3.19	6.83
2	4.55	3.19	4.55
3	4.55	4.55	3.64
4	5.01	5.01	5.01
5	5.01	4.10	4.55
6	3.64	4.10	4.55

TABLE 15.3

Vibration in m/s² at the Plant during Abnormal
Operation

Location	Axial	Vertical	Horizontal
1	3.19	4.55	8.20
2	7.28	9.11	5.92
3	5.92	6.83	XX
4	XX	8.20	9.56
5	10.21	8.69	9.73
6	6.20	8.05	10.4

The fluctuating torque on the mean torque being transmitted by the torsion shaft is responsible for the failure of the torsion shaft in the past. The nature of failure with a slant crack at 45° as seen in Figure 15.19 indicates a clear case of fatigue failure.

For the given dimensions of the present torsion shaft of EN9 material, the mean torsional shear stress induced in the shaft at full load of the motor is around 50 MPa. This stress will increase with the presence of the fluctuating torque due to the fluctuating speed of the electric motor.

Further, the present torsion shaft torsional stiffness was calculated to be about 18069 KN-m/radian and the mass moment of inertia of the 1.492-m long torsion shaft was calculated to about 49.28 kg-m² These values give the first resonant frequency of the torsion shaft at around 96.37 Hz, which is close to twice the synchronous supply frequency of the electric motor. Thus, there is a very strong case of resonant frequency excitation of the torsion shaft, which leads to excessive vibration in the plant. These twice synchronous frequency components are observed in the vibration spectrum near the pinion bearings, as shown in Figure 15.20 and Figure 15.21.

In Figure 15.21 it is also observed that there is a considerable frequency smearing in the spectrum, because of excessive speed fluctuation of the motor in this operational regime.

FIGURE 15.19
Failed torsion shaft indicating fatigue failure.

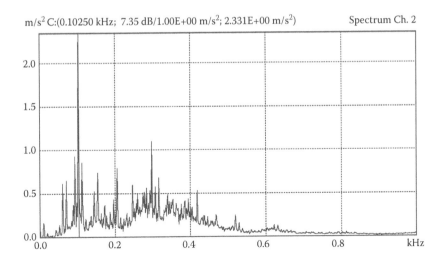

m/s² C:(0.10250 kHz; 7.35 dB/1.00E+00 m/s²; 2.331E+00 m/s²)   Spectrum Ch. 2

FIGURE 15.20
Vertical vibration spectrum showing high amplitudes around 100 Hz at location 5 in normal operation of the plant.

15.3.3 Recommendations

1. The plant must not be run in the present condition. Further running the plant will lead to fatigue failure of the torsion shaft.

2. The motor must be supported with vibration isolators to reduce the transmission of the twice synchronous frequency vibration energy into the plant.

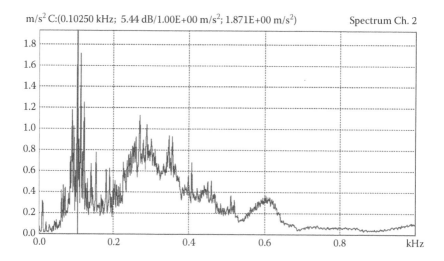

FIGURE 15.21

Vertical vibration spectrum showing high amplitudes around 100 Hz at location 5 during the excessive vibration operation of the plant.

3. The power supply to the motor must be maintained at a steady voltage and frequency.

4. The torsion shaft to be redimensioned to avoid the near twice synchronous frequency resonance condition.

5. Torsional vibration dampers to be put on the torsion shaft system.

15.4 Failure Analysis of a Conveyor System Support Structure

A 300-ton-per-hour (TPH) crushing plant of granite rocks that are used in road construction had a structural failure in the cantilever section of the primary conveyor structure. It was reported that the failure occurred in the structure when the conveyor was not under operation, and the bottom feeder doors to the secondary conveyors were open. Preliminary load study thus indicates that the GR2 member failed because of a downward force on it due to the free fall of the accumulated stock pile on the member. The material of the failed structural members were also tested for their mechanical strength.

In the following sections, the stress analysis of the existing design of the cantilever section is presented using a commercial structural finite element analysis software. Analysis was also done due to the increase in the load on the GR2 member, following increase in the primary storage of the stockpile to 70 kilo-ton. Design recommendations are also provided so that the stock pile can be increased to 70 kilo-ton.

15.4.1 Static Stress Analysis

The existing structure was checked for adequate strength by static stress analysis. As per the drawing specified, the following members were modeled: GR1, GR2, GR3, GR4, GR4, GR5, GR6, and EX1 and EX2. EX1 and EX2 are the extra members provided on top of the cantilever structure. In the FE model, beam elements were used all through the model. As per the IS 808:1989 standard, the following cross sections were selected. The FEM model considered the self-weight of the structural members as well:

- ISMB 300 for GR1 and GR2
- ISMC 200 for GR3, GR4, GR5 and GR6
- ISMC 100 for EX1 and EX2
- Angle of 75 × 75 × 6 for all the internal cross members

The FEM model was constrained at foundation locations for GR1, GR2, and GR6.

The live loads for the conveyor operating at a peak of 450 TPH were calculated, and a load of 3924 N in the vertical direction was estimated. Analysis was also done for the vertical load and lateral load of 3924 N, the results of which are given in Table 15.4.

As per IS 2062:1999, the Grade A FE 410W A was chosen as the material of the truss members. This material has ultimate tensile strength of 410 MPa, and a yield point stress of 230 MPa. The specific gravity of the structural steel material was chosen to be 7.8.

Figures 15.22 and 15.23 show a view of the actual plant.

TABLE 15.4

Axial Loads (N) on the Structural Members

Member	Existing Design	Existing Design with Vertical Load on GR2	Existing Design with Vertical and Lateral Load on GR2	New Design	New Design with Vertical Load on GR2
GR1	−367810	−387070	−340150	−69410	−70732
GR2 bottom	−242100	−261490	−264050	−61507	−55278
GR2 Middle	−112010	−123580	−130800	−124150	−112510
GR2 top	−62236	−68215	−64617	−68525	−62525
GR3	79084	82955	75033	−78112	−71566
GR4	49912	55341	44125	54080	48726
GR5	−75723	−81227	−69992	−79625	−74222
GR6	276560	298480	242750	63543	56556
EX1	99487	108840	83503	107790	98491
EX2	39864	43445	30060	43786	40179

FIGURE 15.22
View of the plant.

FIGURE 15.23
View of the failed GR2.

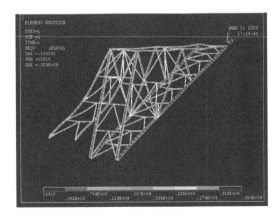

FIGURE 15.24
Von Misses stresses on existing design.

The load analysis on the existing plant indicates that a maximum compressive load of 367810 N exists on GR 1. However, since GR 1 is made of ISMB 300, the maximum compressive load it can carry before buckling is 383730 N. So the margin of safety is very low, and there is a strong possibility of failure in GR1.

The maximum tensile load in the structure is on GR6, of 276560 N, and because GR 6 is made of ISMC 200, it can carry a maximum tensile load of 655500 N, and hence is safe.

The GR 2 is under compression with a loading of 242100 N, and is safe from buckling.

So with the existing design, further increase in load is to be avoided. The loads are summarized in Table 15.4. The von Misses stress in the members of the conveyor support structure obtained from static FE analysis are shown in Figure 15.24.

15.4.2 Material Tests

The material of the failed ISMB and ISMC members were tested at the Materials Testing Laboratory of the Mechanical Engineering Department of the Indian Institute of Technology, Kharagpur. The hardness of the material was determined to be 174 BHN. This material has the correct properties for the structural applications in which it is being used. The yield point stress of the material is 230 MPa.

15.4.3 Additional Load on Existing Design

With extra storage capacity, the member GR 2 becomes covered with the raw material. Estimating a load of 1962 N, at each junction of the member, the static stress analysis was carried out. In addition to the above load on the GR2 member, an analysis was done with an additional lateral load of 3924 N on the joints of GR2.

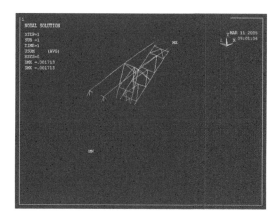

FIGURE 15.25
Nodal displacement of new model.

15.4.4 New Design for Increased Load Capacity

In order to improve on the load and stress analysis, a new design is suggested, where in two stays with cross members are used. The stays are of ISMB 300 section, and attached to GR 2 as shown in Figure 15.25 along with its nodal displacements.

15.4.5 Conclusions and Recommendations

Based on the above stress and load analysis study of the cantilever structure for the crushing and screening plant, the following conclusions and recommendations were made:

1. The existing design is marginally safe. Failure in member GR 1 can occur due to buckling. The cross section of GR 1 can be increased to ISMB 400.

2. Care has to be taken during storage to ensure that the members are not engulfed by the bulk material, for the existing design.

3. However, since the requirement is for storage to a higher capacity of 70 kT, a new design must be adopted, the results of which indicate that the loads induce stress and buckling well below the critical values.

4. In the new design, the new leg and the GR1 can be replaced by ISMB 400.

5. The hinge in the existing design for the GR2 did not have significant effect in the foundation forces.

6. The analysis results summarized in Table 15.4 may be scaled up and down for increasing or decreasing the load on the members, since the FE model is linear.

15.5 Vibration Measurements on a Motor-Multistage Gearbox Drive Set

A vibration measurement audit was done in a newly commissioned material handling plant for a seaport on the motor-multistage gearbox unit. The measurement showed the poor workmanship of the installation and the high level of vibration when the conveyors being driven by the gearbox system were operating under no-load conditions. The measured data are provided in Tables 15.5 and 15.6, and the recommendations on the action to be taken based on the observations are provided for the drive unit. All the vibration measurements were conducted as per the ISO10816-3 standard and in situ calibration of the vibration transducer was done for the measurements. The frequency spectrum of all the vibration measurements showed a predominant peak at the operating speed of the motor at 24.38 Hz.

For each of the drive units, vibration measurements were conducted at 18 locations in three directions (axial, vertical, and horizontal). The vibration measurement locations are provided in Figure 15.26. In addition to the above measurements, vibration measurements only in the vertical directions

TABLE 15.5

Measured Overall RMS Vibration Level [mm/s] (10–1000 Hz) as per ISO 10816-3

Measurement Location	Location Description	Axial Velocity [mm/s]	Vertical Velocity [mm/s]	Horizontal Velocity [mm/s]
1	GBox NDE Left Base	0.821	1.96	0.730
2	GBox Output Left Brg.	0.776	1.46	1.78
3	GBox Intermediate Left Brg.	0.502	1.30	1.88
4	Gbox Input Left Brg.	2.02	1.15	3.03
5	Gbox DE Left Base	1.44	1.59	1.46
6	GBox NDE Right Base	0.913	0.821	1.46
7	GBox Output Brg.	1.01	0.866	3.10
8	GBox Intermediate Right Brg.	0.684	0.866	2.60
9	Gbox Input Right Brg.	1.01	1.15	3.46
10	Gbox DE Right Base	0.577	1.30	0.821
11	Motor DE Right Base	0.548	0.776	1.41
12	Motor NDE Right Base	0.502	0.913	2.46
13	Motor NDE Brg.	1.30	3.24	3.42
14	Motor NDE Left Base	0.577	2.92	1.14
15	Motor DE Left Base	0.721	2.56	1.69
16	Motor DE Brg.	2.10	0.866	2.02
17	GBox DE Brg.	0.866	1.88	3.61
18	Structural Member 2m away from the unit	0.411	1.24	0.721

TABLE 15.6

Vertical Vibration Levels (RMS, 10–1000 Hz) at Foundation Locations of the Machine Unit

Measurement Location	Location Description	On Unit Base (A) [mm/s]	On Structural Base (B) [mm/s]	On Foundation (C) [mm/s]
1	GBox NDE Left Base	1.88	1.73	1.23
5	Gbox DE Left Base	1.59	1.73	1.55
6	GBox NDE Right Base	0.766	0.577	0.548
10	Gbox DE Right Base	1.15	0.821	0.684
11	Motor DE Right Base	0.923	0.639	0.639
12	Motor NDE Right Base	1.60	1.64	1.51
14	Motor NDE Left Base	2.69	2.37	1.98
15	Motor DE Left Base	1.41	2.56	2.68

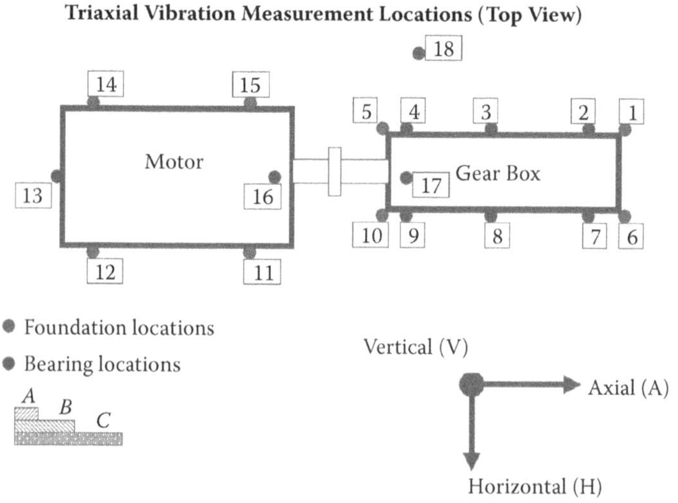

Triaxial Vibration Measurement Locations (Top View)

FIGURE 15.26
Vibration measurement locations.

at the four corner mount locations of the motor and the gearbox were made. At each of the measurement locations, the vibrations were measured at the base of the machine motor-gearbox (Location A), on the structural frame supporting the motor-gearbox (Location B), and then finally on the foundation (Location C). Such measurements show the effect of vibration transmission from/to the motor-gearbox, frame, and the foundation. Figures 15.27 to 15.31 show the different views of the motor–gearbox set with various visible defects and poor workmanship.

Motor current signature analysis was done on the measured motor current from all three phases of the drive motors for the unit by tapping at the

FIGURE 15.27
Oil leaks from gearbox; check/replace seals.

FIGURE 15.28
Nondrive side of motor base not in same level, visible cracks on the base frame; replace base frame.

current transformer (CT) in the motor control cabinet panel. They show a balanced electrical supply, however, with few harmonics in the R phase.

15.5.1 Summary of Observations Made

From the visual inspection, triaxial vibration measurements at 18 locations on and around the motor-gearbox unit, the following general observations were made:

1. There are visible cracks in the base frame of the unit.
2. There are oil leaks in the gearbox.
3. There is corrosion on the drive end (DE) bottom of the motor.
4. The shims have corroded.

FIGURE 15.29
Hopper supported on the edge of the opening, possible failure during operation; recast and install (front view).

FIGURE 15.30
Hopper supported on the edge of the opening, possible failure during operation; recast and install (rear view).

FIGURE 15.31
Corroded shims, thin frame structure; replace.

5. In some locations, the shims have a very high thickness.

6. The jack bolts have corroded.

7. The motor base is not level between the left and right ends from the nondrive end (NDE).

8. The motor is not level at the top between the NDE and the DE as observed from the side.

9. The hopper connected to the gearbox where the raw materials are dropped does not have enough bearing area to take vertical and side thrusts.

10. The base frame structure does not have enough thickness and weight.

11. There is a high vibration level at NDE motor bearing.

12. There is a high vibration level at DE base of the motor.

15.5.2 Conclusions and Recommendations

1. The measurements indicate that at present there is no misalignment between the shafts.

2. Since the NDE motor bearing is being subjected to high vibration levels, and the shafts are aligned, the bearings are being subjected to

fatigue loading, and a premature failure of the NDE motor bearing may occur under load.

3. The unit base has to be made heavier by increasing the channel thickness, and having a larger concrete base, to arrest some of the high vibration levels observed. Usually the foundation weight should be more than two times the weight of the supported unit, however, due to lack of space and heavy structural loads, some flexibility is allowed.

4. In the present situation, if the conveyer unit is loaded, the system will run satisfactorily for some time, but because of the increase in the load, the life of the motor bearings will be reduced, which may lead to failure of the motor.

5. Provisions to arrest the corrosions (in shims, jack bolts, frames, motor casing) observed should be undertaken, since the unit is subject to a saline environment, which may aggravate the situation. Recommend a completely new base for the unit.

6. Though the observed vibration levels are marginally acceptable as per the ISO standards, the fact that this level may rise due to the preexisting condition of the unit has to be checked, by conducting a quarterly vibration level measurement of the unit, since this is a very critical unit to the operation of the port.

7. Periodic motor current signature analysis on the drive motor must be done every quarter.

Bibliography

ASM International. *ASM Handbook Volume 11: Failure Analysis and Prevention*. Materials Park, OH: ASM International, 2002.

Ghosh, N., Y. B. Ravi, A. Patra, S. Mukhopadhyay, S. Paul, A. R. Mohanty, and A. B. Chattopadhyay. "Estimation of Tool Wear during CNC Milling Using Neural Network-Based Sensor Fusion." *Mechanical Systems and Signal Processing* 21, no. 1 (January 2007): 466–479.

Inman, D. J. *Engineering Vibration*. Englewood Cliffs, N.J.: Prentice Hall, 1994.

Kar, C., and A. R. Mohanty. "Monitoring Gear Vibrations through Motor Current Signature Analysis and Wavelet Transform." *Mechanical Systems and Signal Processing* 20, no. 1 (January 2006): 158–187.

Mohanty, A. K. *Fluid Mechanics*, 2nd ed. Englewood Cliffs, N.J.: Prentice Hall, 2010.

Appendix A1: Vibration-Based Machinery Fault Identification

Vibration-Based Machinery Fault Identification Spectral Characteristics

Machine Component Faults	Dominant Direction	Predominant Frequency	Remarks
Shaft/Rotor			
Unbalance	Radial	1 x	Amplitude increases as square of the rotational speed
Misalignment	Axial	2 x	
Looseness	Radial	1/2 x, 0.5 x, 1 x, 1.5 x, 2 x and 2.5 x	Time waveform gets clipped
Rub	Radial	Continuous spectrum	
Crack	Radial	1x, 2x, 3x	Amplitudes decrease with increase of frequency
Bearings			
Journal	Radial	0.42 x to 0.48 x	Due to oil whirl phenomenon
Rolling Element	Radial	Harmonics of CDFs[a] and at 20 to 30 kHz	Detection possible by envelope analysis. High-frequency vibration is due to resonance of bearing components
Gears	Radial	GMFs[b] and sidebands	Cepstrum analysis is preferred to detect families of sidebands
Impellers/Pump	Radial	Sidebands around VPFs[c] and 20 to 30 kHz	High-frequency vibration is due to cavitations
Fan/Blowers	Radial	Sidebands around VPFs	—
Pulley/Belt	Radial	1 and 2 times BF[d]	—
Electrical Motor	Radial	2 times electrical supply frequency	—

[a] Characteristics defect frequencies of bearing components
[b] Gear mesh frequency
[c] Vane pass frequency
[d] Belt frequency

Appendix A2: Vibration Severity Levels

Vibration Severity Levels

RMS Vibration Velocity (mm/s)	Vibration Severity Limits for Machines Measured at Nonrotating Machine Locations (based on ISO 10816)[a]			
	Class I <15 kW	Class II 15–75 kW	Class III >75 kW (Rigid Foundation)	Class IV >75kW (Soft Foundation)
0.28	A	A	A	A
0.45				
0.71				
1.12	B			
1.8		B		
2.8	C		B	
4.5		C		B
7.1	D		C	
11.2		D		C
18.0			D	
28.0				D
45.0				

Note: A = Good, B = Acceptable, C = Monitor closely, D = Unacceptable.

[a] For machines within the rotational speeds of 600 to 12000 RPM, RMS levels are measured within a frequency range from 10 Hz to 1000 Hz. The frequency limits are different for machines with lower operational speeds.

Appendix A3: Field Balancing Techniques

Single Plane Balancing Using Phase and Vibration Measurements

In situ field balancing of rotors with an L/D ratio of less than 0.5 and rotation of less than the first critical speed of the rotor is done by the following procedure:

1. Put a once-per-revolution marker on the rotating disc that can be picked up by an optical photo-tachometer, or use a reluctance-type pickup above a shaft keyway to provide a once-per-revolution trigger signal for rotational speed measurements, as shown in Figure A3.1.

2. Mark the location below the rotational speed transducer on the shaft as a reference point as $0°$.

3. Mount a vibration transducer like an accelerometer on a bearing housing close to the rotor disc. The rotational speed transducer and the vibration transducer can be in the same orientation with $0°$ angle between them as shown in Figure A3.2.

4. For phase angle measurements between the tachometer and the accelerometer, use a dual-channel fast Fourier transform (FFT) analyzer.

5. At the desired rotational speed of balancing, run the rotor and measure the initial unbalance response, V_0 and the phase angle α_0.

6. Attach a trial mass M_T at any location on the disc, and measure the response as V_T and α_t.

7. Calculate the compensation mass, M_c, as $\dfrac{|V_0|}{|V_T|} \times M_T$.

8. Calculate the compensation angle $\alpha_c = \alpha_0 + 180°$.

9. Remove the trial mass and attach the compensation mass at the compensation angle measured from the reference axis.

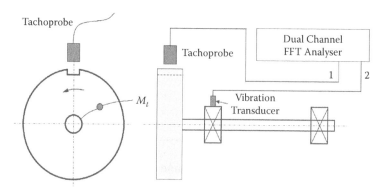

FIGURE A3.1
Tachoprobe used for phase measurements during in situ balancing.

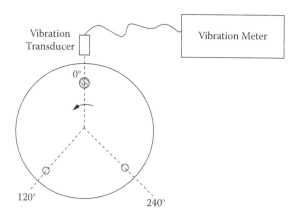

FIGURE A3.2
Vibration transducer location for in situ balancing.

Three-Point Method of Balancing

This method of balancing requires no phase measurements and the same trial weight can be used to estimate the correction mass and location. A total of four vibration measurements are done and the rotor is run at the same rotational speed during the measurements. However, the rotational speed need not be recorded. The trial weight selected should be usually within 10% of the expected unbalance force at the bearings.

1. Mark three locations on the rotor disc at 0°, 120°, and 240°.
2. Mount a vibration transducer with a readout display (vibration meter) on a bearing location close to the unbalanced disc.

3. Measure the vibration level for initial unbalance present without any trial weights as $|V_0|$.

4. Attach a trial weight of mass M_t at the $0°$ location, run the rotor at the desired rotational speed, and measure the vibration level as $|V_1|$.

5. Similarly, after stopping the machine, remove the trial mass from the $0°$ location and place at $120°$ and measure the vibration as $|V_2|$; and at $240°$ as $|V_3|$.

6. Estimate the resultant vector as $V_c = |V_1|\angle 0^0 + |V_2|\angle 120^0 + |V_3|\angle 240^0$, where $V_c = |V_c|\angle\theta_c$.

7. The correction mass is given as $M_c = \dfrac{|V_0|}{|V_c|} \times M_t$.

8. The correction mass is applied on a circle of the same radius where the trial mass was attached at an angle of $\theta_c + 180°$.

Appendix A4: In Situ Alignment by Reverse Indicator Method

Figure A4.1 shows the configuration of the driver and driven machine to be aligned. Dial indicators are mounted to measure the radial displacements (X and Y). The procedure for alignment is as follows:

1. Zero the dial indicators at the twelve o'clock position.
2. Slowly rotate the shaft and bracket arrangement through 90° and record the readings at the three, six, and nine o'clock locations.
3. Calculate the moves in the horizontal and vertical direction at the foundation locations of the driver and driven unit. Apply or remove shims of appropriate thickness at the foundation locations.

$$\text{Movement at location } 1 = \frac{(A+B+C)(X+Y)}{C} - (Y)$$

$$\text{Movement at location } 2 = \frac{(B+C)(X+Y)}{C} - (Y)$$

$$\text{Movement at location } 3 = \frac{(C+D)(X+Y)}{C} - (X)$$

$$\text{Movement at location } 4 = \frac{(C+D+E)(X+Y)}{C} - (X),$$

where X is one half of the driver rim reading difference (from top to bottom or side to side) and Y is one half of the driven rim reading difference (from top to bottom or side to side).

4. Return to the twelve o'clock position to check if the dial indicates zero.
5. Repeat the above steps if required.

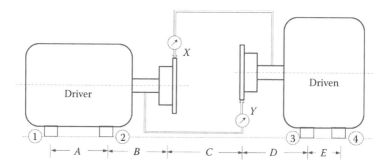

FIGURE A4.1
Location of reverse dial indicators.

Appendix A5: MATLAB® Programs for Basic Signal Processing

1. Sine wave time history

```
clear all
n = 1:1:1000;%1000 number of data points
fs = 1000.0;%sampling frequency of 1000 Hz
delt = 1.0/fs;%sampling time interval
f = 10.0;% frequency of the signal is 10 Hz
t = n*delt;% time in seconds
y = 10.0*sin(2.0*pi*f*t);% 10 Hz sine wave of 10 units
     amplitude
plot(t,y)% time history plot of the sine wave
xlabel('time (s)');% X-axis label
ylabel('Amplitude');% Y-axis label
```

2. Sine wave with random noise time history

```
clear all
n = 1:1:1000;%1000 number of data points
fs = 1000.0;%sampling frequency of 1000 Hz
delt = 1.0/fs;%sampling time interval
f = 10.0;% frequency of first signal is 10 Hz
t = n*delt;% time in seconds
y = 10.0*sin(2.0*pi*f*t) + 10.0*rand(1,1000);%10 Hz sine
     wave signal contaminated with random noise
plot(t,y)% plot of the time signal
xlabel('Time (s)');% X-axis label
ylabel('Amplitude');% Y-axis label
```

3. Square wave-generated time signal

```
clear all
[u,t] = gensig('square',0.1,1,0.001);% 10 Hz square wave
        for 1 second duration sampled at 1000 Hz
ua = 10.0*u;%sqaure wave of 10 units amplitude
plot(t,ua)% plot of the 10 Hz square wave time signal
xlabel('Time (s)');% X-axis label
ylabel('Amplitude');% Y-axis label
```

4. Amplitude-modulated signal

```
clear all
n = 1:1:1000;%1000 number of data points
fs = 1000.0;%sampling frequency of 1000 Hz
delt = 1.0/fs;%sampling time interval
f = 10.0;% frequency of the signal is 10 Hz
fm = 10.0;% 10Hz modulated frequency
fc = 100.0;% 100 Hz carrier frequency
t = n*delt;% time in seconds
y = 10.0*sin(2.0*pi*fm*t).*sin(2.0*pi*fc*t);% amplitude
    modulated signal
plot(t,y)% time history plot of the amplitude modulated
    signal
xlabel('time (s)');% X-axis label
ylabel('Amplitude');% Y-axis label
```

5. Frequency-modulated signal

```
clear all
n = 1:1:1000;%1000 number of data points
fs = 1000.0;%sampling frequency of 1000 Hz
delt = 1.0/fs;%sampling time interval
f = 10.0;% frequency of the signal is 10 Hz
fm = 10.0;% 10Hz modulated frequency
fc = 100.0;% 100 Hz carrier frequency
t = n*delt;% time in seconds
y = 10.0*sin(2.0*pi*fm*t+(sin(2.0*pi*fc*t)));% frequency
    modulated signal
plot(t,y)% time history plot of the frequency modulated
    signal
xlabel('time (s)');% X-axis label
ylabel('Amplitude');% Y-axis label
```

6. Beat signal

```
clear all
n = 1:1:10000;%10000 number of data points
fs = 1000.0;%sampling frequency of 1000 Hz
delt = 1.0/fs;%sampling time interval
f1 = 10.0;% frequency of first signal is 10 Hz
f2 = 11.0;% frequency of second signal is 11 Hz
t = n*delt;% time in seconds
y = 10.0*sin(2.0*pi*f1*t)+10.0*sin(2.0*pi*f2*t);%beat
    signal
plot(t,y)% time history plot of the beat signal
xlabel('time (s)');% X-axis label
ylabel('Amplitude');% Y-axis label
```

7. Fast Fourier transform (FFT) of a signal

```
clear all
n = 1:1:1000;%1000 number of data points
fs = 1000.0;%sampling frequency of 1000 Hz
delt = 1.0/fs;%sampling time interval
f = 77.0;% frequency of first signal is 77 Hz
t = n*delt;% time in seconds
Totaltime = 1000*delt;%length of the time signal
y = 10.0*sin(2.0*pi*f*t);%77 Hz sine wave signal
yfft = 2.0/1000.0*fft(y);% complex FFT of 77 Hz sine wave
      signal
yabs = abs(yfft);%magnitude of the FFT
delf = 1.0/Totaltime;% frequency resolution
n2 = 1:1:500;%FFT results are plotted till N/2 data points
fk = delf.*n2;%frequency values
plot(fk,yabs(1:500))% spectrum of the time signal
xlabel('Frequency (Hz)');% X-axis label
ylabel('Amplitude');% Y-axis label
```

8. Envelope detection

```
clear all
n = 1:1:10000;%10000 number of data points
fs = 1000.0;%sampling frequency of 1000 Hz
delt = 1.0/fs;%sampling time interval
f1 = 10.0;% frequency of first signal is 10 Hz
f2 = 11.0;% frequency of second signal is 11 Hz
t = n*delt;% time in seconds
y = 10.0*sin(2.0*pi*f1*t)+10.0*sin(2.0*pi*f2*t);%beat
      signal
yh = hilbert(y);%hilbert transform to obtain the envelope
      of the time domain signal
yz = abs(yh);% real part of the complex hilbert transform
plot(t,yz)% time history plot of the enveloped signal
xlabel('time (s)');% X-axis label
ylabel('Amplitude');% Y-axis label
```

9. Time domain features of the signal

```
clear all
n = 1:1:1000;%1000 number of data points
fs = 1000.0;%sampling frequency of 1000 Hz
delt = 1.0/fs;%sampling time interval
f = 10.0;% frequency of the signal is 10 Hz
t = n*delt;% time in seconds
y = 5.0+10.0*sin(2.0*pi*f*t);% 10 Hz sine wave with a dc
      offset of 5
ymax = max(y);% Maximum value of the signal
```

```
ymin = min(y);% Minium value of the signal
ym = mean(y);% Mean value of the signal
yvar = var(y);% Variance of the signal
ystd = std(y);% Standard deviation of the signal
ykurt = kurtosis(y);% Kurtosis of the signal
yskew = skewness(y);% Skewness of the signal
```

10. Cepstrum analysis

```
clear all
n = 1:1:1000;%1000 number of data points
fs = 1000.0;%sampling frequency of 1000 Hz
delt = 1.0/fs;%sampling time interval
f = 10.0;% frequency of the signal is 10 Hz
t = n*delt;% time in seconds
y = 10.0*sin(2.0*pi*f*t);% 10 Hz sine wave of 10 units
    amplitude
k = 1:1:1100;% total sample length is for 1.1 seconds
yc = [zeros(1,100) y];%Sine wave delayed by 0.1 seconds
c = cceps(yc);%complex cepstrum
plot(k,c)% cepstrum at quefrencies of 0.1 second and its
    raharmonics
xlabel('quefrency (ms)');% X-axis label
ylabel('Amplitude');% Y-axis label
```

Appendix A6: International Standards on Machinery Condition Monitoring

Standard	Description
ISO 1940-1:2003	Mechanical vibration: Balance quality requirements for rotors in a constant (rigid) state—Part 1: Specification and verification of balance tolerances
ISO 1940-2:1997	Mechanical vibration: Balance quality requirements of rigid rotors—Part 2: Balance errors
ISO 2954:2012	Mechanical vibration of rotating and reciprocating machinery: Requirements for instruments for measuring vibration severity
ISO 4406:1999	Hydraulic fluid power: Fluids—Methods for coding the level of contamination by solid particles
ISO 5348:1998	Mechanical vibration and shock: Mechanical mounting of accelerometers
ISO 7919-3:2009	Mechanical vibration: Evaluation of machine vibration by measurements on rotating shafts
ISO 10816-1:1995	Mechanical vibration: Evaluation of machine vibration by measurements on non-rotating parts—Part 1: General guidelines
ISO 10816-2:2009	Mechanical vibration: Evaluation of machine vibration by measurements on nonrotating parts—Part 2: Land-based steam turbines and generators in excess of 50 MW with normal operating speeds of 1500 r/min, 1800 r/min, 3000 r/min, and 3600 r/min
ISO 10816-3:2009	Mechanical vibration: Evaluation of machine vibration by measurements on non-rotating parts—Part 3: Industrial machines with nominal power above 15 kW and nominal speeds between 120 r/min and 15,000 r/min when measured in situ
ISO 10816-4:2009	Mechanical vibration: Evaluation of machine vibration by measurements on nonrotating parts—Part 4: Gas turbine sets with fluid-film bearings
ISO 10816-5:2000	Mechanical vibration: Evaluation of machine vibration by measurements on nonrotating parts—Part 5: Machine sets in hydraulic power generating and pumping plants
ISO 10816-6:1995	Mechanical vibration: Evaluation of machine vibration by measurements on nonrotating parts—Part 6: Reciprocating machines with power ratings above 100 kW
ISO 10816-7:2009	Mechanical vibration: Evaluation of machine vibration by measurements on nonrotating parts—Part 7: Rotodynamic pumps for industrial applications, including measurements on rotating shafts
ISO 10817-1:1998	Rotating shaft vibration measuring systems: Part 1: Relative and absolute sensing of radial vibration
ISO 11342:1998	Mechanical vibration: Methods and criteria for the mechanical balancing of flexible rotors
ISO 13373-1:2002	Condition monitoring and diagnostics of machines: Vibration condition monitoring—Part 1: General procedures

Continued

Standard	Description
ISO 13373-2:2005	Condition monitoring and diagnostics of machines: Vibration condition monitoring—Part 2: Processing, analysis, and presentation of vibration data
ISO 14695:2003	Industrial fans: Method of measurement of fan vibration
ISO 15242-1:2004	Rolling bearings: Measuring methods for vibration—Part 1: Fundamentals
ISO 15242-2:2004	Rolling bearings: Measuring methods for vibration—Part 2: Radial ball bearings with cylindrical bore and outside surface
ISO 15242-3:2006	Rolling bearings: Measuring methods for vibration—Part 3: Radial spherical and tapered roller bearings with cylindrical bore and outside surface
ISO 16063-1:1998	Methods for the calibration of vibration and shock transducers: Part 1: Basic concepts
ISO 16587:2004	Mechanical vibration and shock: Performance parameters for condition monitoring of structures
ISO 18431-1:2005	Mechanical vibration and shock: Signal processing—Part 1:General introduction
ISO 18431-2:2004	Mechanical vibration and shock: Signal processing—Part 2: Time domain windows for Fourier transform analysis
ISO 18434-1:2008	Condition monitoring and diagnostics of machines: Thermography— Part 1: General procedures
ISO 20958:2013	Condition monitoring and diagnostics of machine systems: Electrical signature analysis of three-phase induction motors
ISO 22096:2007	Condition monitoring and diagnostics of machines: Acoustic emission
ISO 29821-1:2011	Condition monitoring and diagnostics of machines: Ultrasound—Part 1: General guidelines
ISO 18435-1:2009	Industrial automation systems and integration: Diagnostics, capability assessment, and maintenance applications integration—Part 1: Overview and general requirements
ISO 28523:2009	Ships and marine technology: Lubricating and hydraulic oil systems— Guidance for sampling to determine cleanliness and particle contamination
ISO 15549:2008	Nondestructive testing: Eddy current testing—General principles
ISO 22096:2007	Condition monitoring and diagnostics of machines: Acoustic emission
ASTM D445-12	Standard test method for kinematic viscosity of transparent and opaque liquids (and calculation of dynamic viscosity)
ASTM D2270-10e1	Standard practice for calculating viscosity index from kinematic viscosity at 40°C and 100°C
ASTM D1265-11	Standard practice for sampling liquefied petroleum (LP) gases, manual method
ASTM D664-11a	Standard test method for acid number of petroleum products by potentiometric titration
ASTM D4739-11	Standard test method for base number determination by potentiometric hydrochloric acid titration
ASTM D2272-11	Standard test method for oxidation stability of steam turbine oils by rotating pressure vessel
ASTM D1744-13	Standard test method for determination of water in liquid petroleum products by Karl Fischer reagent

Index